# FLOWER GUIDE
# WILD FLOWERS EAST OF THE ROCKIES

*CHESTER A. REED*

2024 Culturea
xte et illustration de couverture : © domaine public (open access licence CC0 1.0 universel)
tion et mise en page : Culturea (Hérault, 34)
rouvez notre catalogue sur http://culturea.fr/
ntact : infos@culturea.fr
primé en Allemagne par Books on Demand Gmbh
sign typographique et layout : Derek Murphy et Reedsy
N : 9791041985807
te de parution : février 2024
livre :
a été composé avec une police open source libre de droits dessinée sur Open Fondry https://open-foundry.com/
est édité en libre access sous licence CC0 (Creative Commons Zero)
fin de conserver l'oeuvre au plus près des caractéristiques du domaine public.
offre la possibilité à son lecteur de partager, dupliquer ou distribuer son contenu, sans restriction ni réserve.
permet de replanter un arbre. SHS Éditions soutient Plantons Pour l'Avenir, fonds de dotation pour le reboisement des forêts françaises
0 projets de reboisement de forêts soutenus depuis 2014 à travers la France https://www.plantonspourlavenir.fr/

## PARTS OF FLOWERS

In order that seed may be set, it is necessary that ripened pollen from the anthers should come in contact with the usually sticky stigma, whence it will be transmitted down the style and germinate the ovule. As is well known to be the case with the higher animals, cross-fertilization is necessary in order to insure a good, vigorous species. It is also evident that should the pollen continue to fertilize the ovule in the same flower, the plants in successive generations might become weakened and finally die out and the species be lost. To avoid such a calamity, flowers are constructed so as to facilitate cross-fertilization and the means that some of them adopt towards insuring that end are remarkable. The stamens on most of the simple flowers usually curve outwards, so that the pollen-laden anthers are far enough removed so that there is little danger of the pollen falling on the stigma, at least until after cross-fertilization has already taken place. Others have either the anthers or stigma ripen first, so that it must necessarily be pollen from another blossom that quickens the seed.

We all marvel at the industry of the honey bee; how tirelessly it buzzes from flower to flower, from each gathering a drop of the nectar, with which it fills its cells; but we do not always realize the double duty it is doing, for it is a most reliable and active agent for the propagations of a great many plants. Many butterflies, bees, and even beetles unconsciously accomplish the same result, and it is now conceded that each has special colors that are attractive to them. For instance, the bumblebee has a strong preference for blues and purples. The observer will also notice that a bee makes the rounds from flower to flower, taking all of one kind and passing by other species. While this habit undoubtedly avoids some complications, even should he mix his drinks and visit in succession flowers of widely different species, confusion would not be apt to result, for the stigma of one species is usually not responsive to pollen brought from blossoms of another family.

Botanists go a step further; not content with the discovery that certain insects like certain colors, they claim (and apparently with good reason) that the bright and showy petals are for the sole purpose of attracting insects; they are, in fact, bill boards advertising the fact that there is a store of honey there ready for the asking. On the other hand, those flowers that are self-fertilized or wind-fertilized have inconspicuous blossoms.

The stigma always partially obstructs the entrance to the food supply, so that the visiting bee must brush against it, and in doing so will leave some of the pollen that he has brought from the last flower visited on its sticky surface. The pollen-dust is attached to the insect in various ways, usually simply by his brushing against the anthers with his hairy body, for it is found that nearly all the useful insects have downy or hairy bodies; other flowers set a sort of spring gun and when the insect steps on the trigger he is showered with the germs (Laurel for example); still others have clefts to catch the legs of visitors, releasing them only if they are strong enough to tear away the pollen masses (such a flower is well illustrated in the Milkweed). Besides having bright colored petals, many of the flowers also have a pleasing odor, this also serving to attract certain kinds of insects; others have very unpleasant odors, like the skunk-cabbage or even like that of putrid meat, as in the carrion flower and the purple trillium, these odors being apparently for the purpose of attracting certain scavenger insects. There are also some flowers, like the evening primrose, that are seen at their best after dusk, when the light-colored petals are widespread and a delicate perfume given off to attract the moths and sphinges that visit them.

It is evident that a flower secreting honey may be visited by unwelcome guests, ones that will accept of the nectar, but will make no useful return. Any insect with a shiny, smooth body whether winged or not, is of little use in fertilizing a plant, for even should it receive pollen, it will in all probability have fallen off before the next flower is visited. Ants being particularly fond of sweet things and so small that they can enter a flower without disturbing the anthers, frequently drain the nectar cups so no useful insect will visit them, and they fail to reproduce their kind. Nature has a number of quite effective ways of preventing thefts of this kind, one of the most common ways being to provide the plant stem with bristly hairs, forming a very difficult barrier for any crawling insect to overcome; others have a tuft of hair at the very entrance to the honey cells which bar the way for unwelcome guests, but readily allow the bee to insert its tongue; still others are protected by recurved leaves or by sticky stems, or as in the toadflax by a two-lipped flower, which will open under the weight of a bumblebee, but is effectively closed to any lighter insect.

The seeds, having been set, are enclosed in a capsule composed of the closed and dried sepals, the petals having fallen off; in a pod as in the peas and beans; or in fruit, as inside an apple, which is formed by the base of the flower enlarging about the seeds with the calyx remaining at the top of the ripened fruit, or in the strawberry, where the seeds are on the outside of the berry and the calyx at the bottom of the fruit. It is plain that should these seeds simply fall to the ground, plants of a single species would soon become so crowded in a small area that the earth could not support them. Consequently various means are furnished different plants for the dispersion of their seeds. A great number, like the thistles, milkweeds and dandelions, have plume-like parachutes provided for each seed, so they can float away on the breeze to new fields; those that have their seeds embedded in fruit are entrusted to birds to be carried where fate wills it; others, like beggar-ticks, burdock, etc., have spines to attach themselves to the clothing of people or to the coats of animals that brush against them.

## KINDS OF LEAVES

As certain insects prey upon plants or rob them of their nectar, so certain plants prey upon insects, literally eating them or absorbing them into their system. Best known among these are the pitcher plants, a swamp species, whose leaves are pitcher-like and with a hood or awning over the top to keep out the rain; these leaves are half filled with a sweet fluid that attracts insects, makes them tipsy and causes their death in the watery grave, the plant feeding largely upon the resulting broth. Of another type is the round-leaved sundew, also a common plant; its leaves are covered with short bristly hairs, a drop of gum glistening at the end of each. A fly investigating these is soon caught in the sticky gum and the leaf slowly folds together, enveloping the victim in what might be termed the stomach of the plant. Perhaps the most interesting and surely the most peculiar plant is the Venus fly-trap, which is found only in eastern North Carolina. At the end of each leaf is apparently a smaller one, perhaps an inch in diameter; this is fringed around the edge and rather bristly in the centre. These central bristles are very sensitive and if touched or an insect lights upon the leaf, the two parts of the leaf instantly clasp together on the central stem as a hinge. If nothing is caught, in a short time the trap opens again; if, however, the attempt has been successful it will remain closed for several days or a week, until the victim is entirely absorbed by the glands on the inner surface of the leaf.

As in the animal world, so in the plant world; always a struggle for existence, the strong surviving and the weak falling by the wayside. The old adage that "In union there is strength" is amply proved by many of the composite flowers, such as the asters and goldenrods, whose stalks are not only capped with numerous flower-heads, but each flower-head is composed of hundreds of little perfect florets, so closely set together that even should an insect but crawl across the flower-head he will fertilize a number of them. That their plan is a good one is seen by the steady increase in the numbers of these flowers and the rapid strides with which they occupy new territory. On the other hand, compare such flowers as the lady's slippers, fringed gentian and numbers of others that are yearly becoming less common.

## FLOWER GUIDE

## WILD FLOWERS EAST OF THE ROCKIES

***Grouped in their Natural Order as in the Latest Edition of Gray's Botany***

### CAT-TAIL FAMILY
### (*Typhaceæ*)

The members of this family are very abundant aquatic herbs with perennial roots. We have two species with differences as noted below. Both have staminate yellow flowers in a spike above pistillate brown ones; the former soon fall or blow away, while the latter develop into the large, familiar, brown cat-tail that is often used for decorative purposes.

(A) Common Cat-tail (*Typha latifolia*) has yellowish staminate flowers encircling the upper end of the flower stalk, and immediately below a long cylindrical mass of brownish pistillate ones. The pollen grains are arranged in fours. Leaves three to eight feet long, sheathing at the base. Found in marshes throughout the United States and southern Canada, flowering in June and July.

(B) Narrow-leaved Cat-tail (*Typha angustifolia*) has narrower leaves, averaging less than ¾ in. broad. The two kinds of flowers are separated by a bare space of stalk and the pollen grains are simple.

## BUR REED FAMILY
## (*Sparganiaceæ*)

The Bur Reeds are marsh-inhabitating plants, some growing along the muddy shores of ponds or streams, while other species are strictly aquatic, growing in the water with floating leaves. Like the Cat-tails they are not in the least dependent upon insects for fertilization. The two kinds of flowers, staminate and pistillate, are always in separate spherical clusters usually alternately arranged along the stem.

(A) Great Bur Reed (*Sparganium eurycarpum*) is stout and erect, two to three feet in height. The mature heads, or fruit, about one inch across; composed of wedge-shaped nutlets arranged in the form of a sphere. The basal leaves are similar to those of the Cat-tail and clasp the stems. These plants are found in the whole of the U.S. and southern Canada, flowering from June to August.

(B) Branching Bur Reed (*S. androcladum*) throws off several weak flower-bearing branches from the angles of the upper leaves.

## WATER PLANTAIN FAMILY
## (*Alismaceæ*)

Genus Arrow-head (*Sagittaria*). Arrow-heads or Sagittarias are among our most beautiful water plants. The leaves vary greatly in shape but are always graceful in appearance. All species have three pure white petals with a golden centre formed by the large anthers. They usually grow in the water but sometimes on the muddy shores, and flower in June.

(A) Broad-leaved Arrow-head (*Sagittaria latifolia*) has broad, arrow-shaped leaves on long petioles from the root. The 3-petalled white flowers grow in whorls of three, the upper ones being staminate and the lower pistillate. Seed, winged on both edges and with a twisted horizontal beak. This species is smooth, but a variety (pubescens) has the stem quite wooly. Common in the whole of our range.

(B) Narrow-leaved Arrow-head (*S. Engelmanniana*) has very narrow leaves with linear sagittate bases. The seeds are winged but the beak points upward instead of being bent at an angle as in the last.

## ARUM FAMILY
## (*Araceæ*)

This is quite a large family of plants containing six genera. All have acrid or pungent juices; flowers closely crowded on a spadix, usually surrounded by a spathe; leaves either simple or compound and of various shapes.

## GENUS (*ARISÆMA*)

Jack-in-the-Pulpit or Indian Turnip (*Arisæma triphyllum*) is the most abundant and the best-known representative of this genus. In moist woods you will find Jack, represented by the spadix, looking out at you from his pulpit, represented by the spathe of the flower. The spathe is light green, more or less striped with brown, especially on the inside; the spadix is also green and has the tiny flowers clustered about its base. The large solid roots are very acrid and fiery to the taste, but are said to have been relished by the Indians. Usually two, thrice-compounded leaves spread shelteringly on long stems over the flower spathe. Large clusters of bright berries remain after the leaves have withered. Flowers throughout U. S. from April to July.

(A) Water Arum (*Calla palustris*) is our only representative of its genus. It is quite a common plant in cool bogs, where it grows from six inches to a foot in height. The beautiful dark green heart-shaped leaves stand above the water on long petioles. A beautiful, waxy-white, spreading spathe is often mistaken for the flower. The true flowers are small and perfect, clustered at the end of a yellow spadix. The flowering season is in June. You may find this plant commonly in cool bogs from N. J. and Mo. northward.

(B) Golden Club (*Orontium aquaticum*) is also our only member of its genus. As you will see by the opposite picture, there is no protective spathe for the golden floral club.

The florets are complete, having six sepals and stamens; they are set closely on the swollen spadix and attract many flies and even water snails that cross-fertilize them simply by crawling over the clubs.

The leaves of the Golden Club are pointed oblong in shape, floating on the surface of the water by means of long stems from the perennial rootstalk. Flowers in May from Mass. to Fla. and westward.

## GENUS (*SYMPLOCARPUS*)

The common Skunk Cabbage (*Symplocarpus fœtidus*) although regarded by many only with disgust, has one claim that cannot be disputed, that of being our first flower to bloom each year. It is not uncommon to find them with the shell-like spathe above ground and the pollen fully ripened even in January, although from the latter part of February to the first of May is the usual flowering season.

The flower spathes show a great diversity of coloring according to their age, ranging from a pale green sparingly streaked with brown to an almost solid purple tone.

The flowers are small, perfect, and closely crowded on the thick fleshy spadix, concealed or partially so by the large, thick purple and green stained hood. The leaves appear after the flower has withered or commenced to do so; they are bright green, large, cabbage-like and strongly veined; quite handsome, in fact. These plants range from N. S. to Minn., and southward, chiefly in boggy ground.

## SPIDERWORT FAMILY
## (*Commelinaceæ*)

(A) Day-flower (*Commelina communis*) is one of a very few of our native plants having pure blue flowers. Each blossom lasts but a single day.

The stem is rather weak, much jointed, and attains heights of one to two feet. Two petals are large, rounded and blue, while the third is tiny and colorless; the whole flower peeps out from a clasping, cordate, heart-shaped leaf or spathe. Found from southern Mass. to Mich. and southward, blooming in rich woods or dooryards from June to Sept.

(B) Spiderwort; Job's Tears (*Tradescantia virginiana*), like the Day Flower, remains open for but part of a day, after which the petals contract into glutinous drops.

The stem is hairy and sticky; from one to two feet high. Three purple petals, three brown, hairy sepals and six orange-tipped stamens compose the flowers. They may be found in rich soil from Me. to Mich. and southward, flowering from June to August.

## PICKEREL-WEED FAMILY
## (*Pontederiaceæ*)

(A) Pickerel-weed (*Pontederia cordata*) is an exceedingly abundant water plant growing profusely in shallow ponds or along the edges of fresh-water streams, and flowering from June to August.

The flowers grow on a spike that proceeds from a small, green, leaf-like spathe; the 3 upper divisions of the 6-parted perianth are partially united, but the 3 lower ones are spreading; they are a light violet-blue with two yellow spots at the base of the upper united parts. A single heart-shaped, cordate leaf clasps the stem about midway, while others on long petioles grow from the rootstalk. Commonly found from N. S. to Manitoba and southward.

(B) Mud Plantain (*Heteranthera reniformis*) has a slender, few-flowered spike proceeding from a small sheath-like spathe. The perianth is blue and regularly 6-parted. The leaves are round-lobed, kidney-shaped, floating on long stalks from the root. Found from Ct. to Neb. and southward.

## LILY FAMILY
## (*Liliaceæ*)

This is an exceedingly large family containing more than 80 species in our range, divided into 33 genera.

(A) Bellwort (*Uvularia perfoliata*) is common in rich woods. The stem, reaching a length of 6 to 18 in., rises from a short rootstalk. A single, straw-colored flower is pendent from the end of each drooping branch; it is long, bell-shaped, and has six narrow divisions. The leaves are light green, lance-shaped, and pierced by the stem. The slightly fragrant flowers are so concealed by their drooping position as to be invisible from above. They blossom in May and June throughout the U. S.

(B) Oakesia; Wild Oats (*Oakesia sessifolia*) has an angular stem from 6 to 14 in. long. The ovate-lanceolate leaves are seated on the stem and not pierced by it. The one or two flowers are similar in size and coloring to those of Bellwort but the interior is smooth while the latter has rough ridges. This species is common from Me. to Minn. and southward.

## ONION; GARLIC
## Genus (*Allium*)

The various species belonging to this genus are very strongly scented, pungent herbs growing from a coated bulb.

(A) Wild Leek; Wild Onion (*Allium tricoccum*) is a woodland plant blooming in May and June. The flowers are in an umbel at the top of a scape 6 to 20 in. high. The flower perianth is divided into six greenish-white sepals. The leaves are oblong-lance-shaped, pointed at both ends, on long petioles from the bulbous root, but usually withering before the flowers appear. Found from N. B. to Minn. and southward.

(B) Wild Garlic (*Allium canadense*) has few purplish 6-parted flowers on slender pedicels from a cluster of bulblets at the top of a scape 10 to 24 in. high. The leaves are grass-like, sheathing the stem above the fibrous bulb. Flowers in May and June in moist meadows, from N. B. to Mich. and southward.

Day Lily (*Hemerocallis fulva*) (European) will, we think, prove a welcome addition to our flora. It is now locally abundant in R. I., Conn., and N. Y. It flourishes best near salt water and spreads rapidly by means of its running roots as well as by seed. The flower stalk is tall, 2 to 5 feet, and at its summit bears eight or nine buds which open one or two a day into large showy flowers.

The perianth is funnel-form, with six spreading orange limbs and six long stamens with large brown anthers. The blossoms appear in July and August, each remaining open for but a single day; this habit makes them very popular for vase flowers as the number of buds on each stalk insures fresh flowers every day for a week or more. The leaves are long and linear, similar to those of the Cat-tail, appearing from a fleshy perennial rootstalk at the base of the tall flower scape.

In the absence of any odor, the beautiful flower cup serves to attract the bees that are necessary for the setting of its seed.

## LILIES
## Genus (*Lilium*)

All the members of this genus are among our most beautiful flowers. In our range it includes eight species, of which seven are natives. The two species of Red Lily can readily be recognized because their perianth, or flower funnel, always opens upward.

Wood Lily; Wild Orange-red Lily (*Lilium philadelphicum*) has a leafy stem 1 to 3 feet high, at its summit bearing one to four erect (not pendulous) flowers; the divisions of the perianth are deep orange-red, lightening in color at the stem-like bases and profusely spotted with dark brown; the outside of the perianth is dull whitish green. The leaves are lanceolate, sharply pointed at each end and whorled about the stem in groups of from three to seven. Its name is rather misleading for, while it is sometimes found in woods, it will be found blooming most profusely in sandy or brush-covered land. Blooms in July and August in sandy soil from N. E. to Mich. and southward.

Turk's-cap Lily (*Lilium superbum*) is a most beautiful plant, prolific in bloom almost beyond belief, sometimes containing from thirty to forty brilliant orange flowers. The bright sepals are always reflexed, sometimes so much so that they remind one of a coiled spring. One has but to touch the large pendent anthers to get a practical demonstration of how the pollen is attached to the body of a bee and carried to another flower, there to be deposited on the sticky stigma of the mature style. Naturally a species so prolific of flower and so capable of being cross-fertilized by foreign agency is in little danger of having its numbers lessened.

The flowers, nodding at the top of a stem ranging from 2 to 7 feet in height, have a six-parted perianth, orange-red, thickly spotted with purplish brown. The lanceolate leaves are crowded along the upper stem and whorled about its lower portion. Blooms abundantly in rich soil, during July and August, from N. B. to Minn. and southward.

Field, Wild, Meadow, Yellow or Canada Lily (*Lilium canadense*) is one of the most abundant of the genus. Imagine a rich meadow, surrounded by deep green woods and covered with thousands of these lilies, their heads hanging and nodding invitingly and seeming fairly to tinkle in the bright sunlight. On the whole, this flower may be regarded as more graceful in form than is the Turk's-cap, but it cannot compare with the latter flower for beauty of coloring. The regular whorled leaves and graceful bending peduncles supporting the hanging "bells" make a conventional design that often appeals to the artistic eye.

The flowers are in terminal clusters of one to twelve blossoms, nodding on long peduncles from the summit of tall leafy stems. The leaves are lanceolate, arranged about the stem at intervals in whorls of three to eight. Flowers during June and July in moist meadows from Quebec to Minn. and southward to Ga. and Mo.

(A) Dog-tooth Violet; Yellow Adder's Tongue (*Erythronium americanum*). These flowers are familiar and welcome ones to all who wander beside woodland brooks in the spring. The name "Violet" is of course a misnomer. It is often locally known as the "Trout Lily" or the "Fawn Lily," both of which names are far more appropriate than those given it generally.

The single 6-parted flower grows at the top of a scape from 5 to 10 in. high. Two elliptical-lanceolate leaves clasp the scape at its base, near the scaly bulb; they are pale green, mottled with purple and white. This species blooms in April and May in moist woods or swamps, from N. B. to Minn. and southward.

(B) Clintonia (*Clintonia borealis*) is a beautiful species, its leaves resembling those of the Lily-of-the-Valley. The three to six pendulous, bell-shaped flowers are cream-colored within and greenish outside. Three large, oblong, pointed leaves clasp the flower scape at its base. The plant is about 6 to 16 in. high; it flowers during June in damp woods from Labrador to Man. and southward.

Wild Spikenard (*Smilacina racemosa*) is quite an imposing plant, with its long, curving, zigzag stem, its many light-green, deeply ribbed leaves and its feathery terminal flower clusters.

The white flowers are tiny but perfect, with a 6-parted perianth, six slender stamens, and a short, thick style. The stem is rather angular and attains a length of from 1 to 3 feet; alternating along it are the large, oval, sharply pointed leaves, with parallel ribs and wavy edge. The perennial rootstalk is thick and fleshy. Spikenard is quite abundant in moist rich, uncleared ground, flowering in May and June, from Me. to Minn. southward.

False Solomon's Seal (*Smilacina stellata*) bears some resemblance to the last species, but the flowers are much larger and few in number, usually only six or eight terminating the zigzag stem. The leaves are broader at the bases and slightly clasp the stem, whereas those of the last species have very short stems. It grows commonly, but not as much so as the last, on moist banks and in meadows, from Me. to Minn. and southward.

(A) Canada Mayflower; False Lily-of-the-Valley (*Maianthemum canadense*) is a very abundant woodland plant, growing in colonies, thousands of them sometimes carpeting pine woods with their dark-green glossy leaves. The two, or three broad, ovate-lanceolate, shiny green leaves are rather heart-shaped at the base, seated on the stem or very nearly so. The flower perianth has four divisions. After the flowering season both of these plants have berries; at first a creamy white, spotted with brown, and later turning to a dull ruby-red.

(B) Three-leaved False Solomon's Seal (*Smilacina trifolia*), the smallest member of the genus Smilacina is found rather commonly in bogs and wet woods. The stem, straight and slender, from 2 to 6 in. high, usually has three leaves, shining green, oblong-pointed, and sheathing at the base, arranged at regular intervals along it. The flowers are white, few in number, on short peduncles in an open raceme and have six petals. It is found from Labrador to Manitoba and southward to N. J. and Mo., flowering during May and June.

Purple Twisted-stalk (*Streptopus roseus*) has, as would be judged from its name, a very angular or twisted stem. At each angle or joint appears an ovate-lanceolate, cordately ribbed, shining green leaf, seated on the stem. From the axils of the terminal leaves appear small flowers on slender, thread-like peduncles; these flowers are sometimes single or, again, in pairs; they have a bell-shaped base and the perianth is divided into six lanceolate, spreading dull purple sepals. The stem, which is rather sparingly bristly hairy, reaches heights of 1 to 2½ feet. This rosy species blooms in May and June in cold moist woods from Newfoundland to Manitoba and southward to the Gulf of Mexico.

Common Twisted-stalk (*Streptopus amplexifolius*) is similar but has greenish-white flowers, the six sepals of which are very strongly reflexed. The plant is somewhat larger, the smooth stem being from 2 to 3 feet in length. It is found throughout northern United States and the southern half of Canada.

Solomon's Seal (*Polygonatum biflorum*) has small greenish, bell-shaped flowers about one half inch in length, hanging in pairs on slender peduncles from the axils of the leaves. The stem is 1 to 2 feet in height. The oblong-lanceolate leaves alternate along, and are partly seated on, the stem; deep green above and glaucous or whitish below. Very common in woods from N. B. to Ont. and southward, flowering from April to June.

These plants receive their name from the thick, fleshy, and knotted rootstalks. They are perennials, each year throwing up new stalks; after flowering these wither away and leave pronounced scars on the roots. These scars suggested the name of Solomon's Seal and the number of them probably accurately denotes the ages of the plants. Both the large and the small species grow in the same localities. They can readily be distinguished by comparison, for commutatum is always larger in all its parts; while it may be but a foot and a half tall it will be stouter and have comparatively larger flowers than its relative. Often it assumes truly gigantic size and may tower above a tall man's head.

(A) Common Green Brier (*Smilax rotundifolia*) is a woody climbing vine with scattered sharp prickles; it climbs by means of pairs of tendrils from the axils of the leaves. Leaves alternating along the stem; round-ovate, sharply pointed at the tip and somewhat heart-shaped at the base. Flowers, few on slender peduncles from the angles of the leaves; perianth bell-shaped, with six short, spreading lobes, pale greenish in color. Common in moist thickets from N. S. to Minn. and southward, flowering in May and June.

(B) Lily-of-the-Valley (*Convallaria majalis*). As a garden flower, this species is probably familiar to nearly everyone. While, as a native, it is only found in some of the southeastern mountain ranges, it is sometimes found in the North as an escape from cultivation. It is a delicately beautiful species, very rich in fragrance and very hardy. The bell-shaped, white flowers grow in a one-sided raceme at the top of a scape, the base of which is sheathed by the two large, broad, oblong-pointed, parallel-veined leaves. It flowers in May and June in the mountains from Va. to S. C.

Indian Cucumber-root (*Medeola virginiana*) is a common woodland plant, but the flowers are so inconspicuous that they are often overlooked; in fact they are often nodding below the upper leaves so as to be invisible. The stem is tall and slender, ranging from 1 to 3 feet in height; it rises from a thick horizontal rootstalk, having a taste similar to that of the cucumber. A whorl of from five to nine ovate-lanceolate, pointed leaves is located midway on the stem; at the top, three smaller but similarly shaped leaves radiate. Above these, or it may be below, because of the curving pedicels, are three flowers. They are pale greenish-yellow; the three sepals and three petals composing the perianth are very much reflexed or curled; they have six stamens each, and one style dividing into three purplish-brown, recurved stigmas.

It is said that the Indians formerly used the roots for food; at the present time they are used for various medicinal preparations. Cucumber-root is found from N. B. to Manitoba and southward to the Gulf, flowering in rich woods during May and June.

## GENUS (*TRILLIUM*)

Trilliums derive their generic name from the fact that all their parts are arranged in threes; three leaves, three petals, three sepals, and a three-parted stigma. The common name of Wake Robin was probably early given because these flowers appear at an early date. As a matter of fact they do not bloom until weeks after the robins have returned to the Northern States. All the purple trilliums have an unpleasant odor resembling that of putrid meat; as they are largely dependent for fertilization upon certain carrion flies, it is very probable that their peculiar color is for the purpose of an added lure for these insects.

Purple Trillium; Birthroot; Ill-scented Wake Robin (*Trillium erectum*) has three purplish-brown petals and three sepals; six stamens exceeding in length the stout spreading stigma. Flower solitary, rising on a short pedicel above the whorl of broad, ovate, pointed, and short petioled leaves. This trillium ranges in height from 6 to 15 inches. It flowers in April and May, in rich woods from Quebec to Ont. and southward.

(A) Large-flowered Trillium (*Trillium grandiflorum*) is the largest of the genus in all respects and is one of the best known and most common species. It can be looked for in any damp, rich woods during May or June. Usually they grow in colonies and it is an exception when one finds a single plant without others being in sight. The stem of this species is from 10 to 18 inches in height; the waxy-white petals are from 1½ to 2 in. in length; as they grow older the color changes to a delicate pink and they curve gracefully backward.

The flower is on a short pedicel above the whorl of broad, ovate-pointed, and short petioled leaves. Found from Vt. to Minn. and southward to N. C. and Mo.

(B) Nodding Trillium (*T. Cernuum*) is quite similar to, but smaller than, the last species. Its blossom is either white or pink and is on a curved pedicel that often bends so as to place the flower beneath the whorl of leaves; the edges of the petals are quite wavy. This demure, bashful little trillium is found from Newfoundland and Man. south to Pa. and Mich.

Painted Trillium (*Trillium undulatum*) has sharply pointed, wavy-edged, waxy-white petals with crimson V-shaped marks at the bases. The ovate leaves are sharply pointed and petioled. It is a common species from Quebec to Ontario and southward.

The Painted Trillium is usually regarded as the most beautiful of the genus. Certainly it is the most abundant. It is more gregarious than others, and we often find large beds of them with their dainty, waxy-white, wavy-edged flowers swaying above the deep green background formed by their broad, whorled leaves. They grow most profusely along the banks of woodland brooks and in cool, moist glens. You will find them most abundant during the latter part of May soon after the wood thrush, that frequents the same locality, makes his appearance from the South. They are always associated in my mind with these birds and with water thrushes that I have often watched as they daintily threaded their way among the numerous plant stalks, entirely concealed above by the numerous leaves, and visible only by placing the head close to the ground.

(A) Star-of-Bethlehem (*Ornithogalum umbellatum*) (European).

The scape, rising from a coated bulb, is from 6 to 12 in. high; at the top is a loose, terminal cluster of from four to eight blossoms. The perianth is divided into six waxy-white sepals, rather greenish on the outside, and with three to seven green nerves; six stamens and a three-sided stigma. The leaves are long, linear, and channeled. Found as an escape, from Me. to Va.

## AMARYLLIS FAMILY
## (*Amaryllidaceæ*)

A family of bulbous and scape-bearing herbs with flat, grass-like leaves and regular six-parted flowers.

(B) Atamasco Lily (*Zephyranthes atamasco*) is an exceedingly beautiful species with pure, waxy-white flowers, only one to a plant, erect at the summit of a scape from 6 to 12 in. high. Perianth funnel-form, with six spreading lobes, a short pistil, and six stamens with large yellow anthers. Leaves long, linear, and channeled. Quite common in moist places or swamps, from Del. to Fla., flowering from April to July.

Yellow Star Grass (*Hypoxis hirsuta*) is the most widely distributed of any of the members of the Amaryllis family. It is very appropriately named. From April until July and more sparingly until September we may see these bright shining golden stars peering at us from a background of green grass. So closely do the leaves of this little plant correspond to the grass leaves, among which they grow, that sharp scrutiny is required to distinguish them. The blossoms are visited by several of the smaller bees for pollen; some of this is often unwittingly carried to the sticky stigma of the next flower visited and cross-fertilization effected.

The flowers are in a loose umbel at the top of a scape from 3 to 8 in. in height; perianth widespread and divided into six shining, golden-yellow sepals, paler and slightly greenish on the outside; the six stamens tipped with large, golden-orange anthers. The slender, narrow, grass-like leaves come from a small bulb together with the flower scape. This species is common from Me. to Manitoba and southward to the Gulf of Mexico.

## IRIS FAMILY
## (*Iridaceæ*)

This family is composed of perennial herbs growing in moist places and having long linear or sword-shaped leaves and large showy flowers. Iris is named from the Greek meaning, rainbow, and it certainly is no misnomer as applied to the Blue Flag or Iris which is the most common of the genus. The perpetuation of this species in healthy condition is insured by the formation of the flower, which is such that self-pollenization is practically impossible. The stamens are directly under the strap-like divisions of the style and the stigma is on the upper surface at the rolled-up tip. Bees are the most frequent visitors.

Larger Blue Flag; Blue Iris; Fleur-de-Lis (*Iris versicolor*). Flower solitary, from a green spathe at the end of a long peduncle; sepals, neither bearded nor crested, but broad, violet, and handsomely veined; petals erect, flat, and spatulate. Leaves sword-shaped, glaucous-green, folded into flat clusters at the base. Very common from Newfoundland to Manitoba and southward, flowering from May to July.

(A) Blue-eyed Grass (*Sisyrinchium angustifolium*), as one would suspect from the name, has grass-like leaves and flowers that make one think of bright little blue eyes as they peep out of the meadow grass in which you will find them.

The Blue-eyed Grasses have recently been separated into thirteen species, differing chiefly in the comparative lengths of the flower spathes, or the lengths of the leaves as compared to the flower stem. The six divisions of the flower are regular, violet, with a yellow or white star-shaped centre; each sepal is blunt, with a thorn-like tip. Common from N. B. to B. C. and southward.

(B) Crested Dwarf Iris (*Iris cristata*). Flowers usually solitary, very delicate in form, and of a light violet color; the sepals have a central crested rib of a bright orange color; the smaller petals are also crested. The tube is long and thread-like. Leaves lanceolate, about 5 to 7 in. long; those forming the spathe are ovate-lanceolate. This attractive little Iris is found on rich wooded hillsides and along streams, from Md. and Ind. southward, flowering in April and May.

## ORCHIS FAMILY
## (*Orchidaceæ*)

This is a large family composed of herbaceous perennials with tuberoid roots or corms. The perianth is composed of six divisions, the three outer being sepals (two of which are often united) and the three inner ones petals, the lower one of which, termed the lip, differs in form from the others.

Yellow Lady's Slipper (*Cypripedium parviflorum*) has usually one, but sometimes three, flowers at the summit of a leafy stem 7 to 20 in. high. The inflated lip is bright yellow, slipper-shaped, and with a rounded orifice open near the base. The two lateral petals are brownish; exceedingly twisted. The broad, bright-green leaves are very prominently ribbed lengthwise, pointed and alternately sheathing the stem. This is one of the northerly species, being found along the northern border of the United States and southern Canada. It grows in colonies and flowers from May to July, in rich woods or bogs.

Showy Lady's Slipper (*Cypripedium hirsutum*) is a magnificent orchid, usually regarded as the most beautiful of the genus. It is of imposing dimensions and has large, fragrant flowers.

The inflated lip is large and balloon-like, about 2 in. in length; white, with crimson-magenta blotches and streaks on the front edge; the sepals are round-ovate and the petals oblong, both pointed and both greenish-white in color. The leafy stem, that bears at its summit the solitary blossom, is from 1 to 2 feet in height. Found locally from Newfoundland to Minn. and southward to Ga. and Mo., flowering in rich woods during June and July.

Small White Lady's Slipper (*Cypripedium candidum*). The flower of this species is of the same size and shape as that of the yellow variety, but the lip is pure white outside and striped with purple inside at the base; the two lateral sepals and the two petals are ovate-lanceolate, greenish, spotted with brown. It is a single-flowered species with numerous leaves. It is found in swamps from N. Y. to Minn. southward.

Pink Lady's Slipper; Moccasin Flower (*Cypripedium acaule*) has solitary flowers surmounting a scape from 8 to 12 in. high; lip large, drooping pink, with a slit in front, instead of a circular opening as in the others. It frequents dry woods and may be found from southern Canada southward.

Although this is the most common of the Lady's Slippers, it is no less beautiful than the others. The flower of the present species is a very ingenious contrivance; it is fertilized by the common bumblebee. The only entrance is through the fissure in the front; it requires considerable pressure to force his burly frame through, but at length he succeeds and the aperture closes behind him. After eating his fill he takes the easiest way out, toward the base where he can see two spots of light. As he forces his way through the narrow passage he comes in contact with a sticky stigma, armed with incurving hairs which remove any pollen he may have on his back; as he continues his struggle out he reaches an anther blocking the passage and waiting to clap its load of pollen on his back.

(A) Green Wood Orchis (*Habenaria clavellata*) has from three to sixteen inconspicuous greenish flowers in a loose spike at the top of a stem from 6 to 18 in. high; lip oblong and with three teeth; spur long, slender, and curved upward and to one side. One or two oblong-lanceolate leaves with obtuse tips clasp the stem near the base while several small bracts alternate along it. Grows in bogs from Newfoundland to Minn. and southward.

(B) Green-fringed Orchis (*Habenaria flava*) is a common green orchis (formerly *virescens*). The lower leaves are oblong-lanceolate, while the upper ones are linear, diminishing in size and passing into the flower bracts. The flower lip is square-ended and toothed; spur slender and about the length of the flower. In the whole U. S. and southern Canada we may find this species growing in bogs.

Habenaria bracteata is similar to *flava,* but the flower bracts are large, being from two to four times the length of the flowers. N. S. to Alaska and south through the U. S.

(A) Yellow-fringed Orchis (*Habenaria ciliaris*) is an attractive and rather common orchis with a tall leafy stem from 12 to 24 in. high. The spike is very closely set with flowers having rounded petals, fringed lips, and slender spurs about an inch in length. The leaves are lanceolate, gradually diminishing in size as they approach the spike and passing into the flower bracts. Found from Me. to Mich. and southward.

(B) Hooker's Orchis (*H. Hookeri*) has a leafless scape from 6 to 12 in. high, at the base of which are two broad, oval, shining, deep-green leaves. The ten to twenty flowers are yellowish-green; lip lanceolate and sharply pointed, less than half an inch long; slender spur about one inch long. Flowers during June and July in woods from Me. to Minn. and south to N.C.

Round-leaved Orchis (*H. orbiculata*) is similar to Hookeri; the lip is oblong, obtuse, and about the same length as the spur. The two basal leaves are almost round. It is common in rich woods from Labrador to Alaska and southward.

(A) Ragged-fringed Orchis (*Habenaria lacera*) does not attract our attention because of its beauty, for its flowers are rather inconspicuous in color. They are, however, remarkable for the peculiarly cut and slashed lip, it being divided apparently with no regard for method or symmetry. The greenish-white flowers are in a dense many-flowered raceme at the summit of a leafy stem from 10 to 20 inches high. The leaves are oblong-lanceolate, diminishing in size to the flower bracts as they reach the raceme. This species is not uncommon in swamps from Newfoundland to Minn. and southward.

(B) White-fringed Orchis (*H. blephariglottis*) has a densely flowered raceme or spike similar to that of the yellow-fringed species, but the flowers are pure white; the lip is not divided but is copiously fringed; lateral sepals rounded, upper ones elliptical and concave; spur nearly an inch long. Leaves lanceolate and gradually diminishing in size as they alternate to the top of the stem. In July and August you may find this species flowering throughout the United States.

(A) Large Purple-fringed Orchis (*Habenaria fimbriata*) is the largest and perhaps the most beautiful of the genus. The pale purple flowers are nearly twice as large as those of the last species; the lip is more deeply fringed. The densely flowered spike is about two inches in diameter and often is twelve inches long. The leafy stem attains heights of from 1 to 5 feet. It is a magnificent plant, the sight of which is well worth the inconveniences necessary to visit its haunts. It grows in swamps throughout the U. S. and southern Canada.

(B) Small Purple-fringed Orchis (*H. psycodes*) has pale purplish flowers in a dense cylindrical spike terminating in a leafy stem, about 1 or 1½ feet tall. The spreading flower-tip is 3-parted and fringed; sepals rounded, petals spatulate and slightly toothed. The leaves are lanceolate and, like those of the fringed orchids, grow smaller as they approach the top of the stem. Flowers in July and August in wet meadows or swamps, from Newfoundland to Manitoba and southward.

(A) Calopogon; Grass Pink (*Calopogon pulchellus*) is an exquisite orchid with a loose raceme of four to twelve delicate pink flowers at the top of a scrape ranging from 6 to 15 in. long. The flowers are apparently upside down as the lip is at the top; it is narrow at the base but broadens into a broad hooked tip, crested on the underside. A single grass-like leaf sheathes the flower scape near its base, as it rises from the solid bulb. It grows in deep swamps and bogs, from Newfoundland to Minn. and south to the Gulf, flowering in June and July.

(B) Arethusa; Indian Pink (*Arethusa bulbosa*) has a solitary magenta-pink blossom topping its slender scape that rises from 5 to 10 in. in height. The petals and sepals are similar in shape and in their proper positions at the top of the flower; the lip rises, then abruptly turns downward, broadens and is adorned with three to five yellow and white crests; margin of lip wavy and sometimes spotted with crimson. From Newfoundland to Minn. and south to Pa. and Mo., Arethusa has been found blooming in swamps during May and June.

(A) Pogonia; Snake-mouth (*Pogonia ophioglossoides*). Snake-mouth is delicate, pure pink in color, and slightly fragrant. Its pollen is not in stemmed masses but is showered on the back of a visiting insect as he backs out of the flower. The stem is from 8 to 13 inches high, bearing at its top a single flower; sepals and petals are similar in shape; the lip is spatulate, prominently crested with yellow and white, and toothed and lacerated. About midway of the flower stem is a single oval leaf and just below the flower is a smaller bract-like one. Pogonia grows in swamps from Newfoundland to Minn. and southward to the Gulf of Mexico, flowering during June and July.

(B) Nodding Pogonia (*P. trianthophora*) has a leafy stem from 2 to 8 inches high. From two to eight small oval leaves alternately clasp the stem; the flowers, which number from one to six, appear singly from the axils of the upper leaves, nodding on slender peduncles; they are small, magenta-pink, and with ovate, three-lobed lips. It is locally distributed from Me. to Wisc. and southward.

(A) Whorled Pogonia (*Pogonia verticillata*) has a single flower on a long stem, 8 to 12 in. high; the sepals are greenish-yellow, long, linear, with the edges rolled or folded together; the petals are oblong-lanceolate and purple; the lip is also purple, wedge-shaped, three-lobed and with a hairy crest, down the middle. Five lanceolate and stemless leaves are in a whorl about the stem just below the flower. It is a peculiar, inconspicuous plant found locally in moist woods from Me. to Wisc. and southward.

(B) Showy Orchis (*Orchis spectabilis*) is a charming early-blooming orchid found in flower from April to June in moist woods, often under hemlock trees. Two broad, ovate, deeply ribbed, beautiful leaves sheath the flower scape at its base. The four to twelve flowers are loosely racemed at the top of the scape which is from 5 to 10 in. high. The magenta-pink petals and sepals are united to form a hood; the lip, curving abruptly downward, is broadly ovate and white; each flower has a short spur and is bracted. This species is found throughout the U.S.

(A) Rattlesnake Plantain (*Epipactis pubescens*) is a common orchid having beautiful leaves, radiating from the fleshy, creeping rootstalk. The scape is 6 to 15 in. high and carries at its top densely flowered sepals and petals united to form a hood. It is found in the whole of the U. S., flowering in July and August.

(B) Ladies Tresses (*Spiranthes cernua*) is so named because of the braided arrangement of its flowers. The leaves are few, grass-like, sheathing the scape near its base. The scape is 6 to 15 in. high, has several small bracts, and ends in a 2- or 3-ranked spiral raceme of white or creamy flowers; petals and upper sepal joined, lateral sepals lanceolate; lip ovate-oblong with a rough tip. Common in moist fields or woods from Me. to Minn. and southward.

Slender Ladies Tresses (*S. gracilis*) is slender, has its flowers in a single-ranked 1-sided or slightly twisted raceme; lip green, with a white wrinkled margin. Leaves small, ovate basal. Found in dry ground from N. S. to Manitoba and southward.

(A) Heart-leaved Twayblade (*Listera cordata*) belongs to a genus containing five species.

Like most of the orchids, they are largely or wholly dependent upon insect aid for fertilization. The weight or shock of an alighting insect on the broad lip causes a small gland within the flower to rupture and cover the pollen, just below, with a sticky fluid that causes it to adhere to the head or body of the insect and thus be transferred to the next flower.

The stem of this species is from 3 to 10 in. high. At the top is a few-flowered raceme; the sepals and petals are similar and spreading; the lip is drooping, longer, two-cleft and madder-purple in color. This species flowers during June and July in swampy woods from N. J. to Colo. and northward to the Arctic coast.

(B) Twayblade (*Liparis lilifolia*) although having the same common name, is of a different genus. It is a more attractive plant, having two broad basal leaves and larger flowers with a broad ovate lip. It grows in woodland from Me. to Minn. and southward.

## BIRTHWORT FAMILY
## (*Aristolochiaceæ*)

A small family of low herbs or twining vines, with but two genera and few species.

Wild Ginger (*Asarum canadense*) may be found flowering in rich woods during April and May, from Me. to Mich. and southward. It has two large, heart-shaped leaves on long petioles from the base; deep green above and lighter below, soft, wooly, and handsomely veined.

The leaves are very beautiful, but it is the solitary flower that makes this plant so interesting. Small, dully colored, on a weak, short stem that barely raises it above ground and often leaves it concealed by the dead leaves that carpet the woods in early spring.

The flower is bell-shaped, with three short, sharply-pointed spreading lobes; six stamens with short anthers and a thick style with six radiating stigmas. Another species (*grandiflorum*), found in Va. and N. C., has but one leaf and flowers twice as large, or two inches in length.

Pipe Vine; Dutchman's Pipe (*Aristolochia macrophylla*). The Dutchman's Pipe is chiefly a southern plant or vine, being found from Pa. and Minn. southward. It has a woody, climbing stem that may attain lengths of from 10 to 40 feet. The very large, deep-green, veiny leaves that alternate along the stem are very beautiful. In the dull, greenish-yellow flowers, however, lies the chief interest of the botanist. Its stigma matures and withers away before the ripening of the pollen, thus making the plant dependent upon insects for its perpetuation.

The throat is filled with tiny hairs, all pointing inward, so ingress is easy but egress impossible. Entering insects are held prisoners, living upon the nectar, until the stigma withers and pollen ripens; after this the hairs in the throat lose their rigidity and the pollen-dusted and well-fed prisoners are allowed to escape. Their memories are poor or the pollen feast is well worth the imprisonment, for they usually immediately hie to another blossom and force their way in, of course pollenizing the flower in so doing.

## BUCKWHEAT FAMILY
## (*Polygonaceæ*)

This family is divided into seven genera and many of these are further divided. They are all inconspicuous in flower. The genus Rumex, to which our common Sorrels belong, contain seventeen species: that of Polygonum, which contains the Knot-weeds, has 32 species included in its six sub-genera.

(A) Lady's Thumb; Persicaria; Knotgrass (*Polygonum persicaria*) (European). This is a very common weed everywhere in damp places, especially about farmhouses. The small, crimson-pink flowers are in dense spikes terminating the branching stems that are from 1 to 3 feet high. The lanceolate pointed leaves, that alternate along the angled and sheathed stem, are rather rough and usually have a dark triangular spot in the middle.

(B) Common Smartweed; Water Pepper (*P. hydropiper*) has similar shaped flowers of a greenish color. The leaves are lanceolate and very acrid. It is very abundant in wet places throughout our range.

## PINK FAMILY
## (*Caryophyllaceæ*)

(A) Common Chickweed (*Stellaria media*) (European). Although this is an introduced weed, so hardy and prolific is it that probably it now exceeds in numbers any of our indigenous plants. It grows profusely about dooryards and along roadsides everywhere. The corolla consists of five white, very deeply cleft petals, and the calyx of the same number of larger and longer green sepals. The leaves are ovate, small, opposite, on small stems about the length of the leaves. The plant stem is either simple or branched and ranges from 2 to 10 in. in height.

(B) Long-leaved Stitchwort (*S. longifolia*) has larger flowers than the last, but the petals are very narrow and so deeply cleft as to appear to be ten in number instead of five. The sepals are nearly but not quite as long as the petals. The stem is weak and usually supported by surrounding grasses or vegetation. The leaves are small, linear, and pointed at both ends. Common everywhere in wet places.

(A) Corn Cockle (*Agrostemma githago*) (European). The Corn Cockle is very closely related to the Campions (genus Lychnis). It is an annual with an erect and rather downy stem; it branches but slightly, each branch being terminated by one or two large handsome magenta flowers with an expanse of one to two inches. The calyx is densely hairy, as are also the lanceolate leaves that grow oppositely on the stem. We find it as an escape from gardens or in waste places near grain fields.

(B) Ragged Robin (*Lychnis Flos-cuculi*) (European). This species, which is also known as Meadow Lychnis, is noteworthy because of the slashed appearance of its five crimson petals. The flower calyx is deeply ribbed and is of a brownish-purple color, as is also the upper part of the flower stem; both are sticky and hairy. It is sometimes found in waste land or moist places where it has escaped from cultivation.

(A) Bladder Campion (*Silene latifolia*) (European). We have several Campions, some natives and some introduced. The present species was brought to us from Europe. It has very unusual blossoms, in that the calyx is very inflated, almost globular and handsomely marked with darker green, so as to often give it a very similar appearance to that of the citron melon. The five white petals are cleft in twain for nearly their whole length.

It is a common escape from gardens and may be found blooming from June to August along roads or in dry waste places from Quebec to Minn. and south to Va. and Mo.

(B) Evening Lychnis; White Campion (*Lychnis alba*) (European). This is another attractive species introduced from Europe. The petals are white, deeply cleft, and crowned at the base with little petal-like divisions; the calyx is inflated and often deep pink on the ribs. The leaves are smooth edged and oppositely on the stem that grows from one to two feet high. Escaped from gardens, from Me. to N. J. and west to Ohio.

Bouncing Bet (*Saponaria officinalis*) (European). This is probably the most hardy and the most widely distributed of our adventive members of the Pink family. It increases very rapidly by means of underground runners as well as by seed. It is very commonly known as "Soapwort," because of the fact that the mucilaginous juice from the crushed leaves will form a lather if they are shaken in water; it is said that it was, in olden days, used for washing purposes.

The plant stem is quite stout, smooth, erect, and sparingly or not at all branched. At the top is a corymbed or flat-topped cluster containing many flowers; petals, notched or sometimes quite deeply cleft, and with an appendage at the top of the long claws that, bent at right angles, enter the long, tubular, veined, greenish, 5-notched calyx.

From July until September Soapwort blooms profusely in waste places along railroad beds and beside dusty roads where few other flowers are able to flourish. It was one of the first of foreign flowers to be introduced into this country.

(A) Maiden Pink (*Dianthus deltoides*) (European). A handsome rose-colored Pink that has become naturalized along the Atlantic Coast and is quite abundant in some localities, in fields and waste places. The flowers grow singly, or in pairs, at the ends of the branching stem; the petals are broad, wedge-shaped, and finely toothed.

(B) Fire Pink; Catchfly (*Silene virginica*) is one of our most brilliantly colored wild flowers, the petals being either deep crimson or scarlet; the five petals are oblong, 2-cleft, long-limbed, and five in number. The lower leaves are thin and spatulate, the upper ones oblong-lanceolate. Both stem, leaves, and calyx are rather hairy. This species is found in open woods from southern N. J., western N. Y., and Mich. southward.

Wild Pink (*Silene pennsylvanica*) is another beautiful native species, with bright pink flowers and a low, sticky stem; the upper leaves are small, and the numerous basal ones lance-shaped. It is rather common from Me. to N. Y. and southward.

## PURSLANE FAMILY
## (*Portulacaceæ*)

(A) Spring Beauty (*Claytonia virginica*), although very delicate in appearance, is among our earliest flowering plants.

The weak stem is usually very crooked and is often prostrate on the ground; two linear-lanceolate leaves clasp it oppositely about halfway up. The opened flowers, somewhat less than an inch across, have five petals, two sepals, and five golden stamens that mature before the stigma. It is found in moist woods from Me. to Mich. and south to the Gulf.

(B) Purslane (*Portulaca oleracea*) (European) has a prostrate, juicy stem and thick, fleshy leaves; the latter are wedge-shaped with rounded ends. The stem is very branching and spreads or radiates from the root. The flowers are tiny, solitary, and yellowish, seated in the whorls of leaves that terminate the branches. Found in waste places anywhere and possibly indigenous in the Southwest.

## WATER LILY FAMILY
## (*Nymphæceæ*)

(A) Cow Lily; Yellow Pond Lily (*Nymphæa advena*) is not unattractive and is interesting in its makeup. The leaves are thick, rough, ovate, slit or lobed to the stem, which is long and hollow. The flower is raised above the surface of the water on a long, hollow stem. What appear to be six large green and yellow petals are in reality sepals; the real petals are numerous, stamen-like, inserted with the very numerous stamens under the golden-yellow rayed disk that forms the stigma. Very common in still or stagnant water.

(B) Water Lily; Water Nymph (*Castalia odorata*) needs no introduction to our readers. To my mind, it leads all other flowers in beauty, grace, purity, and fragrance. It is composed of four sepals, greenish on the outside and whitish within, and numerous pure, waxy-white petals. They sometimes are gigantic in size, often spreading five or six inches across. It flowers from June to Sept. in ponds or slow-moving water.

## RANUNCULUS FAMILY
## (*Ranunculaceæ*)

(A) Water Plantain (*Ranunculus laxicaulis*) is a rather common marsh-inhabiting Buttercup, with five to seven narrow yellow petals. The stem is stout but rather weak and angled, at each joint sending out a clasping lanceolate, almost toothless leaf. The flowers, which are about ¾ in. broad, are on long peduncles terminating the branching stem that rises from 1 to 2½ feet. It is found in bogs, ditches, and muddy places from Me. to Minn. and south to the Gulf.

(B) Marsh Marigold (*Caltha palustris*) is the very common marsh herb usually, but erroneously, called "Cowslip." Its leaves are very commonly used and marketed for food. The flowers are perfect, have no petals but from five to nine (usually the former) golden-yellow, shining sepals, and numerous brighter stamens. The stems are hollow and furrowed. The leaves are round, kidney-shaped, usually with scalloped edges. Marsh Marigold is abundant in swamps or wet meadows from Newfoundland to Alaska and southward through the United States, flowering in April and May.

(A) Creeping Buttercup (*Ranunculus repens*) is, as per its name, a creeping plant. The stem is prostrate, creeping along the ground and striking new roots from the junctions of the leaf and flower stems with the main one. The flowers are large and broad-petalled, both the petals and stamens being a deep shining golden yellow. This species is indigenous in the West, but probably introduced from Europe in the East, where it is found chiefly near the coast, in ditches or along the edges of marshes.

(B) Common Buttercup; Crowfoot (*R. acris*) (European). Even though we have quantities of native Buttercups, it is this handsome foreigner that is the most abundant; this is the species that is found in fields everywhere, the one that delights the little folks and figures in many of their childish games.

The leaves and stem of the Crowfoots are very acrid, but not poisonous; on this account they are shunned by cattle and horses. This accounts in part for their abundance in most fields and pastures.

(A) Tall Meadow Rue (*Thalictrum polygamum*) is one of the characteristic plants of swamps and edges of streams. Should its neighboring plants be three or four feet high, we find the plume-like flowers of this species triumphantly waving above them on stems of five, six, or even seven feet tall.

The stalk is rather stout and grooved, pale-green, stained with maroon. The long-stemmed leaves are many times compounded into small, lobed leaflets of a pale, dull, blue-green color. The flowers are in feathery clusters; each individual flower having numerous white filaments, no petals, but usually four or five early-falling sepals.

From June to September we may find the mist-like flowers of Meadow Rue in swamps, from Labrador to Manitoba and south through the United States.

(B) Pasque Flower (*Anemone patens*) has a solitary erect flower with five to seven purplish sepals. Leaves divided and cut into narrow, acute lobes. Both stem and leaves covered with silky hairs. This species is found on prairies from Wis. and Montana southward.

(A) Wood Anemone; Wind Flower (*Anemone quinquefolia*). The stem is slender and from 4 to 8 in. high. Three leaves radiate from a point about two thirds up; each on a long stem and divided into three to five, toothed, ovate leaflets, The solitary flower rises on a slender peduncle from the junction of these three leaves with the stem proper. It has four to seven sepals, most often five; white inside and purplish white on their outer surface. The flower has an expanse of slightly less than one inch, but is rarely seen fully expanded. The Wind Flower is common in woods or thickets from Nova Scotia to the Rockies and southward.

(B) Rue Anemone (*Anemonella thalictroides*) has four to nine sepals (usually six), numerous orange-tipped stamens and a broad stigma. There are several flowers on exceedingly slender peduncles, rising from the whorl of leaves. The latter are on slender stems, have heart-shaped bases and three-lobed ends; rather small, pale-green above and with a whitish bloom below. It is found in the same localities and the same range as the last species, with which it associates.

(A) Purple Virgin's Bower (*Clematis verticillaris*) is probably the most rare species of Clematis. It grows in rocky, hilly, or mountainous woods, most abundantly in northern portions of its range, which is from Quebec to Hudson Bay and south locally to Del. and Pa. It is a climbing woody vine, supporting itself by the bending or clasping of the leaf stalks. The flowers grow singly, on long stems from the axils of the leaves or from the end of the vine. They are large and handsome, the four thin, purple, pointed, translucent sepals spreading from two to four inches when fully expanded. The leaves are divided into three leaflets, ovate, pointed, with a heart-shaped base.

(B) Virgin's Bower (*Clematis virginiana*) is a beautiful, graceful, climbing, twining vine found throughout our range. The small greenish-white flowers, with four or five petals, grow in clusters from the leaf axils; staminate and pistillate ones are on separate plants. In fall, the beautiful silky plumes of the seed pods gives this species the name of "Old Man's Beard."

(A) Wild Columbine (*Aquilegia canadensis*) is one of our typical, early woodland plants, graceful in form and beautiful in flower. It grows in rocky woodland throughout our range, flowering from April to June.

The stem is very slender, wiry, and graceful, quite branching, and attaining heights of one to two feet. The flowers are heavy, which causes them to nod from their slender, thread-like peduncles. A quantity of nectar is secreted in the base of each red spur, serving to attract butterflies, moths, and often the Ruby-throated Hummingbirds, for those birds are very partial to red colors.

(B) Goldthread (*Coptis trifolia*) is a small woodland plant receiving its name from the slender, thread-like, golden-yellow roots. These roots are characteristic and readily identify the species. The leaves are evergreen, deeply shining green in color, 3-parted and notched, on long petioles from the root. The white flower has five or six early-falling sepals; it is usually solitary on a scape from 3 to 6 in. high. Common in rich woods throughout U. S. and Canada.

(A) Monkshood; Aconite (*Aconitum uncinatum*) is an attractive wild flower with a slender, rather weak stem often supporting itself against other species. The flowers are quite large and handsome. The five sepals are very unequal in size and shape; the upper one large and hood-like, concealing two small petals within it. The leaves are firm, three- to five-lobed and notched, on slender petioles. In rich, moist woods from Pa. southward, flowering from June to September.

(B) Hepatica; Liverwort (*Hepatica triloba*). If we except the Skunk Cabbage, the beautiful Hepatica is the first of our flowers to appear. Its stems are thickly covered with fuzzy hairs; the three-lobed, smooth-edged leaves are rather thick and coarse, lasting through the winter but turning a ruddy color, while the new ones, that appear with the buds, are light-green and radiate above the older prostrate ones. A single blossom appears at the end of each long fuzzy scape; it is about one inch broad, and has five to ten pale-purple or lilac sepals.

Hepaticas bloom from March to May in open woods from N. S. to Manitoba and southward.

Mandrake; May Apple (*Podophyllum peltatum*) belongs to the Barberry Family (*Berberidaceæ*), a small family of shrubs or herbs, divided into five genera of but one or two species each. The present species is quite common in rich woods, or in shady moist ground, from western N. E. to Minn. and southward flowering in May. The bare stalk rises to heights of 10 to 12 inches, then branches into two long-stemmed, light-green, large, spreading leaves; the latter are five- to nine-parted, lobed, notched, and unevenly balanced. From the forked joint of the leaves hangs a solitary white flower on a short, slender, curving peduncle, this is very delicate, nearly two inches across, and of six petals and twice as many stamens.

The fruit is large and lemon-shaped, yellow in color, ripening in July. It is the fruit that gives it the name of May Apple. While the leaves and stem are poisonous, the fruit is not, but has a peculiar, acid, sickish flavor.

(A) Bloodroot (*Sanguinaria canadensis*). Closely following on the heels of our handsome Hepatica we find the delicate flowered Bloodroot unfurling its leaves and expanding its flowers in rich, rocky, open woodland. The flowers are very delicate; the petals stay but two or three days anyway, and a breath of wind may blow them off sooner.

After the flower is gone, the leaf develops rapidly and becomes very large and imposing, with many divisions and lobes. The root is reddish and is filled with a bloodlike juice, as is also the stem. Bloodroot is common from N. S. to Minn. and southward. It flowers in April and May.

(B) Prickly Poppy (*Argemone mexicana*) is a handsome Mexican plant found in the southwestern portions of the United States. It has a prickly stem from one to two feet high. The stemless leaves have sharp lobes, also armed with prickles. The flower is bright yellow, has four petals, and numerous orange-tipped stamens. The flowers give no nectar but plenty of pollen to the bees that visit them.

Celandine (*Chelidonium majus*) (European) is abundant almost everywhere in the eastern half of our country.

The stem is quite stout and very branching; at the end of each branch is a loose cluster of buds on slender pedicels. These open one or two at a time, so that the plant keeps in bloom for a long time; in fact, the flowering season extends from early in May to the end of September. The flowers are half an inch or more broad, with four golden-yellow petals, a slender, pointed green pistil, and numerous yellow stamens. The seed-pod is long and slender.

The thin, soft leaves are very handsomely divided into three-to seven-lobed leaflets. Both stem and leaves have a bright yellow, very acrid juice, that stains everything it comes in contact with. Celandine is often known in Europe as "Swallow-wort" as it is supposed to commence flowering with the coming of the swallows and to cease with their departure. Its generic name also originated in this belief.

(A) Dutchman's Breeches (*Dicentra Cucullaria*). This peculiarly flowered herb belongs to the Fumitory family, a family of delicate, smooth plants with watery juices and compound, dissected leaves.

The flower stalk, proceeding from the root, attains heights of from 5 to 9 inches and bears a loose raceme of four to eight white, inverted flowers; the four petals are united in pairs, two of them forming a large double-spurred sac, and the other two very small petals forming a protection for the stigma. The double sac is white, stained with yellow. The leaves are on long petioles from the rootstalk; they are pale sage-green in color, 3-parted and finely slashed. Dutchman's Breeches may be found blooming in April and May in rich, hilly woods from N.S. to Minn. and south to N.C. and Mo.

(B) Squirrel Corn (*D. canadensis*) is similar, but the white, sac-like petals are stained with purple, the spurs are shorter and rounder, and the flower is slightly fragrant. The roots have little tuberous appendages resembling grains of corn. This species is found in the same range as the last.

## MUSTARD FAMILY
## (*Cruciferæ*)

(A) Toothwort; Crinkleroot (*Dentaria diphylla*). During the latter part of April or in May we will find white, crosslike flowers of Toothwort often growing side by side with Anemones. Its stem is stout and smooth, and rises to heights of 8 to 12 inches. Two 3-parted, notched-edged leaves with short stems are set oppositely on the flowering stalk, above the middle; other larger, similar ones are on long petioles from the rootstalk. Its root is crinkled and with toothlike appendages. It is found in rich woods from N. S. to Minn. and southward.

(B) Whitlow Grass (*Draba verna*) (European) is a weed that we find along roadsides, waste places, or barren fields. The flowers are small, and the four white petals are deeply notched. The scape is from 1 to 5 in. high. The leaves are all basal, lance-shaped, and lobed or toothed.

(A) Common Mustard (*Brassica nigra*) (European) is extensively cultivated in Europe for the small dark-brown seeds that form a valuable article of commerce, being used for the table condiment and for various medicinal purposes.

In our country Mustard is regarded as a pest; it is a very strong, hardy plant, soon overrunning sections where it gets a foothold. The stem is very branching and grows to heights of from 2 to 7 feet. The four-petalled, light-yellow flowers are in small dense clusters at the ends of the branches; a trail of small, erect seedpods is left in the wake of the flowers as they continue to bloom along the lengthening stem. The leaves have a large, terminal notched lobe and smaller lateral ones.

(B) Hedge Mustard (*Sisymbrium officinale*) (European). This common weed has tiny, four-petalled yellow flowers that bloom all summer, along the lengthening stem, and leave numerous tiny pods closely set against the stem. The leaves are more angular and more finely divided than those of the Common Mustard.

## PITCHER PLANT FAMILY
## (*Sarraceniaceæ*)

A small family of bog-inhabiting plants having hollow pitcher-formed or trumpet-shaped leaves.

Pitcher Plant; Huntsman's Cup (*Sarracenia purpurea*). Few plants are as little known generally as this species. It is one of the most interesting ones that we have. The shapes of both the leaves and blossoms are clearly shown in the opposite picture. The pitchers, or basal leaves, may number from three to a dozen, all radiating from the root and all with the orifice up. An examination shows that each pitcher is partially filled with water. Just below the rim of the leaf, on the inside, is a sticky substance to attract insects; as these enter, they pass downward over countless little hairs, all pointing downward. These make it very difficult for insects to crawl out of the pitcher, and many of them become exhausted and are drowned in the water. As these insects decompose, they are absorbed by the plant.

The Pitcher Plant is local in bogs from Labrador to Manitoba and southward.

## SUNDEW FAMILY
## (*Droseraceæ*)

(A) Thread-leaved Sundew (*Drosera filiformis*) has long, linear film-like, erect, very hairy leaves. The flowers are numerous and loosely racemed at the top of a slender smooth scape; they have five small purple petals, five stamens, and several 2-parted stigmas. This species is found in wet sandy soil from New England to Delaware.

(B) Round-leaved Sundew (*Drosera rotundifolia*) is one of the most common of the Sundews, it is found in moist, sandy, or peaty soil from Labrador to Alaska and south to Pa. and Cal. The leaves are numerous, quite round, and on long stems from the root. The leaves are thickly covered with hairy glands that exude drops of a clear glutinous fluid. These dew-like drops deceive insects into alighting on the leaves. Having caught a victim, the leaf slowly folds about it and digests it.

The flower stalk of this species grows from 5 to 9 in. high, is reddish colored, and often has one or two branches at the top. The one to twenty-five flowers that it has during the flowering season are white.

## SAXIFRAGE FAMILY
## (*Saxifragaceæ*)

(A) Grass of Parnassus (*Parnassia caroliniana*) is a pretty little swamp or meadow plant growing from 8 to 24 inches high. The flowers are a delicate creamy white, finely veined with greenish, and borne singly on long scapes; a single heart-shaped leaf clasps each flower scape a short distance above its base. The basal leaves are long-stemmed, rather thick and coarse in texture, smooth-edged and bluntly pointed.

We find this species in bloom from the latter part of June until the end of September, most abundantly in the latter month. It ranges from Newfoundland to Manitoba, south to Va. and Mo.

(B) Early Saxifrage (*Saxifraga virginiensis*) is a tiny-flowered plant that loves dry, sunny, rocky hillsides. It flowers during March and April. The leaves are all basal; spatulate in shape, blunt ended, either rough-edged or toothed, rather coarse in texture, narrowing toward their base into clasping stems. Saxifrage is common from N. B. to Minn., south to Ga. and Tenn.

Mitrewort (*Mitella nuda*) has a few flowers, very short-stemmed, in a loose raceme at the top of a leafless, slightly hairy scape 4 to 7 in. high. The flowers have five petals, each with the edge beautifully fringed so as to give the flower a crystalline appearance almost like a snowflake. We find this species from Labrador to Saskatchewan, south to Ct. and Mich.

(A) Two-leaved Mitrewort (*Mitella diphylla*) is a larger and sturdier species with similar flowers, but with two very short-stemmed, heart-shaped leaves clasping the flower stem oppositely about halfway up its length. Found in rich woods from N. E. to Minn., south to N. C. and Mo.

(B) Foam Flower; False Mitrewort (*Tiarella cordifolia*) has the general appearance of the last species. The slender, hairy flower scape, rising 6 to 12 in. from the rootstalk, has at the top a loose panicle of many small flowers, each on a long, slender stem, thus differing from the short-stemmed flowers of Mitella. Foam Flower is common from N. S. to Minn. southward, flowering in May and June.

## ROSE FAMILY
## (*Rosaceæ*)

(A) Meadowsweet (*Spiræa salicifolia*) is a common and beautiful shrub that grows along the edges of woods, swamps, or even roadsides. Its handsome pyramidal clusters of flowers are in evidence during July and August. The stem is straight, slender, woody, and yellowish buff; along it, at close intervals, alternate the lanceolate, toothed, short-stemmed leaves. At the top is a spire-like panicle of fleecy flower clusters. Each flower has five round, white petals and numerous long, pink stamens that give the flowers a feathery appearance and a rosy tint. It ranges from N. Y. to Mo. and southward.

(B) Hardhack; Steeplebush (*Spiræa tomentosa*) is one of our most beautiful flowering shrubs. The flower spike is more slender and steeple-like than that of Meadowsweet, and the flowers are a beautiful shade of pink. The leaves are more closely alternated and are dark green above and lighter below. Steeplebush grows in low ground from N. B. to Minn. and southward.

(A) Wild Strawberry (*Fragraria virginiana*). The hairy stems of both leaves and flowers rise directly from the running rootstalk. The flowers, several of which grow on each stem, are wheel-shaped, have five rounded white petals, and narrow lanceolate greenish sepals. After the flowering season, the green center expands, becomes pulpy, and finally turns red on the outer surface; the numerous seeds are in little pits provided for them on the surface of the berry.

The Wild Strawberry is common in fields and pastures throughout our range.

(B) High Bush Blackberry (*Rubus allegheniensis*) is a tall branching shrub with slender brown stems, from three to ten feet long, armed with stout, slightly recurved prickles. It is from this species that the well-known variety was developed. The leaves are divided into three to five ovate, pointed, toothed leaflets, with a ribbed and hairy surface. The flowers have five green sepals alternating with the narrow white petals. This species is very common everywhere.

(A) Common Cinquefoil; Five-finger (*Potentilla canadensis*). This species is the most common of the Five-fingers. It is often mistaken for the Wild Strawberry, because of a similarity between the leaves of the two species, although those of this species have five divisions while those of the Strawberry have but three. The flowers are shaped like those of the Strawberry, but have bright-yellow petals. It is very common in the United States and southern Canada.

(B) Silvery Cinquefoil (*Potentilla argentea*) is a common and very handsome species found in dry, barren ground throughout our range, but most abundantly near the coast. It is smaller than the preceding, being from 5 to 12 in. high. The little yellow flowers are clustered at the ends of the branches. The stems and the undersides of the divided and deeply cut leaves are covered with fine, white, silvery wool, contrasting sharply with the dark green of the upper surfaces. This species bloom from May until September.

(A) Agrimony (*Agrimonia gryposepala*) is a common weed found on the borders of swamps or thickets. It has a tall, hairy, simple stem from two to four feet high.

The flowers are in a long, many-flowered spike at the top of the stalk. Each flower is tiny, has five yellow petals, and numerous orange stamens, giving the spike a bright, golden-yellow appearance. It is a common plant from N. B. to N. C. and westward to Cal.

(B) Marsh Five-finger; Purple Cinquefoil (*Potentilla palustris*) is in character quite like the foregoing species. It is the only one, however, having purple flowers, and is easily recognized on that account. The flowers are nearly an inch broad, larger than those of the other Cinquefoils.

The stem grows from 6 to 20 inches long and is rather woody at the base. Purple Cinquefoil grows in swamps or cool bogs, from Labrador to Alaska and south to N. J., Pa., Ia., and Cal., flowering during July and August.

(A) Creeping Dalibarda (*Dalibarda repens*) is a delicate woodland plant, found from N. B. to Manitoba and south to N. J., Ohio, and Mich. It has creeping, densely tufted rootstalks, from which spring numerous heart-shaped leaves on long petioles; these leaves, their stems, and the flower stalks are downy, the former being scallop-edged or toothed.

Dalibarda has two kinds of flowers: The first on long, upright scapes spread about half an inch, have five oval white petals and many stamens. The second are cleistogamous ones (fertilized in the bud) on short curving peduncles from the root. These last flowers are fertile, while many of those with petals are not. Dalibarda blooms from June to September in rich woods.

(B) Yellow Avens (*Geum strictum*) grows in moist locations in swamps or thickets. The texture of the whole plant, leaves and stems, is rough and coarse. The root leaves are interruptedly pinnate, the segments being wedge-shaped and toothed. The flowers have quite large golden-yellow petals and a downy receptacle. This species is common from Newfoundland to Manitoba and south to N. C. and Mo.

Swamp Rose (*Rosa carolina*). Wild Roses are very common throughout our range and, of course, are familiar to every one. The Swamp Rose is a very bushy species, growing from one to nine feet high. It is very common on the edges of swamps or streams, and in low ground, throughout our range.

The flowers are two or three inches broad and have numerous yellow stamens radiating from the greenish-white centre. The stem of the Swamp Rose is sparingly armed with stout, wide-based, curved thorns.

Pasture Rose (*Rosa humulis*) is the most abundant of all our Wild Roses and grows in profusion in all dry, rocky places. It does not grow as high as the Swamp Rose, rarely exceeding three feet in height, but the slender stems are more branching and often grow in large, tangled masses. The flowers are about the same size as those of the Swamp Rose, but are usually solitary at the ends of the branches.

The stem is armed with straight, slender, light brown thorns or prickles, two of which are set oppositely on the stem at its junctions with the leaf stems.

Sweetbrier; Eglantine (*Rosa rubiginosa*) is a very beautiful species of Wild Rose introduced from Europe. We may find it blooming quite commonly in dry, rocky pastures and waste places during June and July. It is remarkable for and easily identified by the sweet-scented aromatic fragrance of its leaves. The stems are long and arching, growing from 2 to 6 feet in height; they are brown and are armed at frequent intervals with short, decidedly recurved thorns or prickles.

At regular intervals along the stem are close-set, compact clusters of flowers and leaves. The leaves are made up of five or seven very small leaflets, rounded-ovate in form and with the edge finely double-toothed, and covered beneath with fine, sticky, glandular hairs. The flowers are also quite small, especially when compared to the very common Pasture and Swamp Roses, being only from 1 to 2 inches in diameter. Eglantine is found from Nova Scotia to Michigan and southward to Virginia and Tenn.

## PULSE FAMILY
## (*Leguminosæ*)

(A) Wild or Blue Lupine (*Lupinus perennis*) receives its generic name from the Latin of wolf, because it was thought that the species preyed upon the soil and made it infertile for other kinds of plants. It is a very common species in sandy places and we often see it on the banks along railroads. The stem is quite stout, erect, hairy, and branching. The leaves have long, slender stems; the leaf, proper, is palmately divided into seven to eleven narrow, smooth-edged leaflets.

The flowers are in long, showy, terminal spikes of pea-like blossoms. Lupine is very common through the United States, east of the Rocky Mountains.

(B) Blue False Indigo (*Baptisia australis*) is a tall branching species with a stem from 3 to 6 feet in height. The leaves are divided into three spatulate-shaped leaflets. The violet-blue flowers grow in long loose spikes; they are about one inch long, have a four- or five-toothed calyx, straight keel and wings, and short standard. The seed-pod has a spur at its tip. This species is common from Pa. to Ga. and west to Mo.

(A) Wild Indigo (*Baptisia tinctoria*) is a very branchy and very bushy herb. The stem divides soon after it leaves the ground, the slender branchlets extending equally in all directions. The leaves are three-parted, wedge-shaped, dull green with a white bloom that gives them a bluish-green appearance. The yellow, butterfly-shaped flowers are in loose clusters at the ends of all the branches.

The roots of Wild Indigo are used by drug concerns for the compounding of a number of medicines. An indigo dye, of a poor quality, can also be made from the plant. Wild Indigo grows in dry, sandy soil from Maine to Minnesota, flowering from June to September.

(B) Rattlebox (*Crotolaria sagittalis*) receives its name because the seeds rattle about in the large, inflated, blackish seed-pod. It is an annual herb, with a hairy-bending stem and stemless, toothless, pointed-oval leaves alternating along it. The yellow, pea-like flowers are in small clusters at the ends of the branches. It is found in sandy soil, chiefly along the coast, from Mass. to Fla. and Texas and, in the Mississippi basin, to Indiana and South Dakota.

(A) Goats Rue; Cat Gut (*Tephrosia virginiana*). We find this herb in most all dry, sandy, waste places from N. H. to Minn. and southward.

It is a pea-like plant with a simple, silky-haired, erect stem, leafy to the top where it terminates in a dense raceme or panicle of yellowish-white flowers marked with purple. The flowers are large and numerous; they have a rounded standard but little longer than the wings and keel. Its roots are long, very slender, and very tough.

(B) Partridge Pea (*Cassia Chamæcrista*) is a handsome species with large, showy yellow flowers measuring about 1¼ inches across; often the five, large, rounded petals have purplish spots at their bases; after flowering long, erect seed-pods are left in the place of each of the blossoms.

The leaves of the Partridge Pea are long and compounded of 20-30 small, blunt, lance-shaped leaflets, each with a tiny awl-like point. The stem is erect, rather smooth, and grows one or two feet tall. We find this plant in dry or sandy fields throughout the United States.

(A) Rabbit-foot Clover; Stone Clover (*Trifolium arvense*) (European). The stalk of this species is soft, silky, and from 4 to 10 inches high. The light-green leaves have three leaflets with blunt tips. The flower-heads are composed of numerous florets; it is the long, pink, feathery tips of the five-parted calyx that gives the blossom its silky fuzziness; it is quite fragrant and is visited by the smallest butterflies. You may find this species everywhere within our range.

(B) Red Clover (*Trifolium pratense*) is the most common and the most valuable species of clover. One would hardly believe, knowing how abundant it is in all parts of our range, that this Clover could have been introduced and have become so widely distributed, yet such is the case. One reason that it does so well in this country is that we have a very large number of bumblebees, and it has been found that clover is so dependent upon these insects for fertilization, that, without them, it will soon die out.

The little florets, composing the globular flower-head, are bright crimson-pink. The three leaflets that make up each leaf have whitish-green triangles in the middle.

(A) Alsike or Alsatian Clover (*Trifolium hybridum*) (European) is quite similar to our native White Clover, but the stem is stout, branching, and juicy. The trifoliate leaves of this species are unmarked and have a simple, rounded end, not notched, but the edge of the leaf is very finely toothed. The florets composing the round flower-heads are cream-colored, tinged with pink; they are very fragrant and laden with nectar.

(B) White Clover (*Trifolium repens*) is the most common of the White Clovers. It is supposed to be indigenous in the northern parts of our range. It is highly prized as forage for cattle and is often cultivated in fields for that purpose. It is also a favorite with keepers of bees.

Its stems are smooth, reclining, and 4 to 10 inches long. The leaves are composed of three leaflets, heart-shaped or notched at the ends, and usually with a more or less distinct triangular mark in the middle. The flowers are creamy white, slightly pinkish, and very fragrant.

(A) Yellow Clover; Hop Clover (*Trifolium agrarium*) (European). This Clover is very common in the eastern half of the United States and southern Canada; we find it growing most abundantly along roadsides and in dry or sandy fields.

The stem is quite smooth, slender, erect, and slightly branching and grows from 6 to 15 inches high.

The flower-head is oblong, densely crowded with small, golden-yellow florets, having an alternate scaly arrangement. They bloom from the bottom of the head upward, and, as they mature, turn yellowish-brown and are reflexed, resembling dried hops. Its flowering season is from June until September.

(B) Yellow Melilot; Yellow Sweet Clover (*Melilotus officinalis*) (European) is a common, weed-like plant, found everywhere in waste places. The stem is tall and branching, growing from 2 to 4 feet high. The leaves are trifoliate, each leaflet being finely toothed and the middle one having a short stem with a double bend. The yellow, clover-like florets are in long, loose racemes, terminating the branches; they have a sweet fragrance.

(A) Alfalfa; Lucerne (*Medicago sativa*) (European) is found growing wild in waste places or fields most anywhere in our range. It makes an excellent fodder for cattle and will grow in waste, sandy places where it is impossible to raise crops of hay.

The stalk is smooth, slender, branching, and erect; it grows from 1 to 2 feet high. The leaves are three-parted, on long, slender stems with narrow stipules at their base. The purple flowers grow in short, loose racemes at the ends of the slender branches; the seed-pod is curiously twisted or coiled.

(B) Cow Vetch; Blue Vetch (*Vicia Cracca*) is a trailing herb with a weak, angled stem; it is common on the borders of thickets or the edges of cultivated fields. The stem grows from 2 to 3 feet long and climbs over grasses or low brush by means of small, slender tendrils at the ends of the leaves.

The compound leaves are made up of twenty to thirty small, oval leaflets, each tipped with a tiny sharp-pointed bristle. The light violet-colored, bean-like flowers grow in one-sided racemes.

(A) Ground Nut; Wild Bean (*Apios tuberosa*) is an exceedingly beautiful climbing vine, attaining lengths of 4 or 5 feet, crawling over walls or fences, or twisting itself about shrubs or other plants. Its pear-shaped, tuberous root is edible, as every country boy knows.

The leaves of the Ground Nut are compounded of five, or sometimes seven, ovate-pointed leaflets; they are toothless, smooth, and light green. The flowers grow in dense, rounded clusters on slender stalks from between the angles of the leaves and the plant stem, and are maroon or lilac-brown. We find Ground Nut in bloom during August and September in damp ground, usually on the borders of swamps or wet meadows, from N. B. to Minn. and southward to the Gulf.

(B) Wild or Hog Peanut (*Amphicarpa monoica*)is a dainty, trailing vine 2 to 7 feet long. The delicate, light-green leaves are thrice compounded, on slender stems from the angles of which are small, drooping clusters of magenta-lilac blossoms. Other fruitful blossoms at the base of the plant develop into pear-shaped pods with single large seeds.

## FLAX FAMILY
## (*Linaceæ*)

(A) Common Flax (*Linum usutatissimum*) (European) This slender species is more attractive than the last because of its larger flowers. The stem is very slender, from one to two feet in height, and each of its few branches is terminated with one or two delicate, violet-blue flowers; these measure about three quarters of an inch broad, or slightly more.

This is the species that is cultivated very extensively in Europe, and less so in this country, for its linen fibre and its seed oil, both of which have a very extensive commercial use.

(B) Wild Yellow Flax (*Linum virginianum*) is a slender perennial species with a smooth stem from 1 to 2 feet in height. The flowers have a calyx divided into five sepals, a corolla of five petals, five stamens, and pistils, perfect and symmetrical flowers fertilized by small bees and bee-like flies. The small leaves are thin and have but one rib. This species may be found in dry woodland from Me. to Minn. and southward.

## WOOD SORREL FAMILY
## (*Oxalidaceæ*)

(A) White Wood Sorrel (*Oxalis Acetosella*) is one of the most delicate and dainty of our woodland flowers. It is commonly found in cool, damp situations and is very partial to mountainous regions. The flowers are very frail looking, about an inch broad, borne on long, slender peduncles from the root. The leaves are also on long, slender petioles from the root; they are trifoliate or clover-like, each of the three leaflets being inversely heart-shaped—that is, with the end notched and with two rounded lobes. White Wood Sorrel is found from N. S. to Saskatchewan and south to N. E., N. Y., and in mountains to N. C.

(B) Violet Wood Sorrel (*Oxalis violacea*). The long, slender flower stalks bear at their summits three or more pale magenta flowers, similar, except in color, but a trifle smaller than those of the white species.

The leaves of both these Sorrels are very sensitive and fold up if handled; they also close at dusk and open in the morning.

Yellow Wood Sorrel; Lady's Sorrel (*Oxalis corniculata*) is not a woodland plant but is very common along roadsides, in gardens, dooryards, and fields. The pale-green, slender stem is quite erect, branches but little, if at all, and grows from three to twelve inches tall. The leaves are long-stemmed and trifoliate, the three leaflets being broadly heart-shaped. They are very sensitive and close if roughly handled.

The leaves have very acid and sour juices, similar in taste to those of the common Red Sorrel that, by the way, belong to an entirely different family (Buckwheat). Country school children often chew the leaves of both of these, as the sour taste has an agreeable twang.

The bright golden-yellow flowers are quite fragrant; they open only in the sunshine and close tightly at night. They grow in few-flowered umbels at the end of the stem on slender peduncles from the axils of some of the leaves. After their flowering season little erect, pointed pods take the place of the flowers. This species is a very common herb or weed throughout our range.

(A) Wild Geranium; Cranesbill (*Geranium masculatum*) is one of our most common woodland plants, flowering from May to July. The stem, the leaves, and the flower calyx are rough-hairy, the former being quite stout and branching and attaining heights of 1 to 2 feet. The large magenta or pale-purple flowers are in loose, few-flowered clusters at the ends of the branches; the petals are large and rounded and slightly overlap.

The leaves of the Geranium are very coarse and fuzzy, and the surface is often spotted with white or brown; they are palmately divided into five lobes, each of which is sharply toothed and pointed. It is very common from Me. to Manitoba and southward.

(B) Herb Robert (*Geranium Robertianum*) is a smaller edition of the last. Its flowers are similar, but smaller and coarser in texture. Its leaves are smaller and usually more deeply cleft. The stem is usually stained with red; both this and the leaves emit a strong odor when bruised.

Herb Robert is common from Me. to Minn. and southward.

## MILKWORT FAMILY
## (*Polygalaceæ*)

(A) Fringed Polygala (*Polygala paucifolia*) is a dainty and low perennial, usually rising but four or five inches from the ground; the stem bends sharply as it enters the soil and continues into a long, slender rootstalk often a foot in length. Either one or two flowers are at the summit of the stem; they are quite large, being nearly an inch in length; the two lateral sepals are large and wing-shaped; the three petals are joined together to form a tube, through which the yellow stamens and pistil protrude. Polygala is common in damp, rich woods from N. S. to Manitoba and southward to the Gulf, flowering during May and June.

(B) Milkwort (*Polygala polygama*) is a slender-stemmed species from 5 to 15 inches high; the stem is closely crowded, alternately, with narrow, oval, pointed, stemless leaves. The dull crimson flowers are borne in long, slender racemes at the top of the stem. It is quite common everywhere in dry, sandy soil.

(A) Field or Purple Milkwort (*Polygala sanguinea*) is a sturdy little pink-headed plant that grows in fields or meadows or along roadsides.

The flowers, proper, are concealed beneath the large, broad, scale-like, crimson-pink sepals that tightly overlap each other and form the head; these scale-like sepals correspond to the wings on the Fringed Polygala, the true petals and minutely crested keel being shorter and not visible from the outside. The small, stiff, acutely pointed leaves are densely alternated on the stem up to the flower-head. The plant grows from 6 to 12 inches high, and abounds throughout the U. S.

(B) Cross-leaved Milkwort (*Polygala cruciata*) has spatulate-shaped leaves arranged in fours around the stem—cross-like. The stem is quite branchy, and grows from 4 to 14 inches high. At the end of each branch, seated within the four terminating leaves, is a dainty little globular pink flower-head.

We find this species around the edges of swamps or in rather moist fields, from Me. to Minn. and southward to the Gulf of Mexico, flowering from June until September.

## SPURGE FAMILY
## (*Euphorbiaceæ*)

Snow-on-the-Mountain; White-edged Spurge (*Euphorbia marginata*) is a large, bushy herb often cultivated because of its beautiful, white-margined foliage. The stem is very stout and branchy, and grows from 2 to 3 feet high. The leaves are dark green, large, ovate-pointed, and seated on the stem; the lower ones are quite similar in shape to those of the common Milkweed and are alternated on the stem; those near the end of the branches are crowded, opposite, or whorled about the stem; the terminal ones have the edges of the leaves more or less widely margined with white.

The flowers are rather small, grouped in clusters in the centre of the terminal cluster of margined leaves. The staminate and pistillate flowers are on different plants. The involucre is five-parted and has five white petals.

When broken both the leaves and stems exude quantities of a milky juice. This species of Spurge grows in dry soil from Minn. and Ohio west to Colorado, and is sometimes found in parts of the East.

## JEWEL-WEED FAMILY
## (*Balsaminaceæ*)

Jewel-weed; Spotted Touch-me-not (*Impatiens biflora*) is a common rank-growing herb with a stout but fragile branching stem. The large, inflated flower-sac, which is really one of the three sepals, is orange-yellow, spotted with brown. Two of these singular flowers droop from the ends of each thread-like peduncle, but only one flowers at a time.

The slim seed-pod is the cause of two very commonly applied names—Touch-me-not and Snapweed. When nearly ripe these pods can scarcely be touched but what they will suddenly, almost explosively, burst and scatter their seeds in all directions. One not acquainted with their ways is always startled when he accidentally brushes against the mature Touch-me-not.

The leaves are very delicate in appearance, and their light, slender stems are almost translucent; they are ovate, round-toothed, dull-green above, and whitish-green below. Common in shady places throughout the United States.

## MALLOW FAMILY
## (*Malvaceæ*)

Common Mallow; Cheeses (*Malva rotundifolia*) (European) is a very common weed about dooryards, especially in the country, and along the edges of cultivated fields. The long stalks spring from biennial roots and creep over the ground, the branches being 6 to 24 inches in length. The dark-green, round leaves are very handsome; they have a shallow-lobed and very firmly toothed edge and are deeply, palmately ribbed. The leaves, their stems and the plant stems are rather rough.

The small, widespread, bell-shaped flowers are clustered close to the stalk on short stems from the axils of the leaves. The five petals have notched tips, are white, delicately tinted with pink or pale magenta, and have veinings of a deeper shade. The seed is hard, flat, and rounded, composed of a dozen or more carpels; it is eaten by children with great relish, these being the "cheeses" that give the species one of its common names.

(A) High Mallow (*Malva sylvestris*) (European) is a tall biennial with a coarse branching stem, often attaining a height of 3 feet, or even more on waste land. Both the stems and the leaves have a thick covering of hair. The flowers grow in clusters of perhaps a half-dozen from the axils of the leaves; they have five heart-shaped petals of a purplish color, with two or three conspicuous veins of a darker shade.

The Mallows get their generic name of Malva, in allusion to the soothing effect of the mucilaginous juices of the root and stem.

(B) Musk Mallow (*Malva moschata*) (European) is a similar species with the leaves deeply and palmately slashed and toothed. Several hairy branching stems proceed from the perennial root to heights of 1 or 2 feet. The flowers are peculiar in that the ends of each of the five rose-colored petals are roughly notched, looking as though they had been bitten off.

This species received its name from the fact that when the leaves are crushed they give forth a slight odor of musk. It is quite abundant in northern New England and southern Canada.

Rose Mallow (*Hibiscus Moscheutos*) is a tall, leafy perennial, bearing flowers that easily rank as being among the largest and most beautiful of any of our wild flowers. The stem is quite stout and inclined to be hairy. The large leaves are ovate-pointed and toothed; they are stemmed, and alternate along the main plant stalk. The lower ones are often three-lobed.

The flowers grow on short stems at the end of the upright stalk. But one usually blooms at a time and there are not a great many buds; what they lack in profusion of bloom this species fully makes up in size, for its blossoms measure 4 to 6 inches across. The five large petals are a delicate rose color, conspicuously veined, and often with crimson bases. The long, slender pistil divides at the tip into five flat-headed stigmas; for more than half its length it is encased in the long stamen column, the sides of which are covered with yellow anthers.

The Rose Mallow grows in swamps and marshes near the coast, from Mass. southward, and along the shores of the Great Lakes to Mich. It blooms from July to September.

## ST. JOHNSWORT FAMILY
## (*Hypericaceæ*)

Common St. Johnswort (*Hypericum perforatum*) (European) is a wanderer from the Old World that, having reached our hospitable shores, proceeded to multiply and overrun the native plants so that it is now regarded by farmers as a pest along with the Wild Carrot and Mustard. If it is true that in the struggle for existence the fittest survive, then surely this species must be one of the fittest; we often see it growing lustily in circumstances under which few plants could exist. It grows promiscuously in fields or along roadsides. Even a generous sprinkling of tarvia, received when the roads were sprinkled, failed to kill this plant, although many other species died from the effects.

It has a slender but tough stem from 1 to 2 feet high; it has numerous short branches, each crowded with tiny, stiff, oval leaves. The upper branches terminate in clusters of five-parted, golden-yellow flowers with numerous long yellow stamens. This species blooms from July until September.

(A) Marsh St. Johnswort (*Hypericum virginicum*). The stem is slender, erect, and from 1 to 2 feet in height, growing from perennial running rootstalks. The comparatively large leaves are light green with brownish spots and a white bloom on the underside; they are closely set, oppositely, on the stem.

The flowers are in small clusters terminating the branches; the five petals are of a pinkish, flesh color and surround three groups of golden-yellow stamens. Common in moist places.

(B) St. Johnswort (*Hypericum ellipticum*) has a simple usually four-angled stem, sometimes with a single branch near the top. It grows from 8 to 20 inches high and is quite leafy. The leaves are comparatively large, about the same shape as those of Marsh St. Johnswort. At the top are a few five-petalled flowers with bright golden-yellow stamens. The ovoid pods succeeding the flowers are brownish. This species is found in damp places or along streams from Me. to Minn. and south to Pa.

## ROCKROSE FAMILY
## (*Cistaceæ*)

(A) Frostweed: Rockrose (*Helianthemum canadense*). This little perennial is very remarkable and unique, because late in autumn crystals of ice form about the cracked bark of the root. It is also remarkable for the fact that it has two sets of flowers, the first ones in June and later ones in July or August.

The leaves are small, oblong-lanceolate, hoary with white hairs on the underside, alternating along the stem that rises from 10 to 18 inches high. Both the early and late flowers are fertile. Frostweed grows in sandy, dry soil from Me. to Minn. and southward.

(B) Hudsonia (*Hudsonia tomentosa*) is a low-branching, little shrub rising only 5 or 10 inches above ground. Its branching stems are closely crowded with tiny, scale-like oval leaves about one half inch long.

The small yellow flowers that are crowded along the ends of the branches open only in sunshine; the five tiny yellow petals surround numerous stamens and a long, slender style. Hudsonia is found on sandy shores from N. B. to Va., and along the Great Lakes.

## VIOLET FAMILY
## (*Violaceæ*)

(A) Bird-foot Violet (*Viola pedata*) is a well-known and very characteristic Violet. The flowers of this species are the largest of the blue Violets; they are blue-violet or purple-violet and have a bright orange centre, formed by the large anthers.

The leaves grow on long petioles, in dense tufts, from the root; each leaf is cut into five to eleven parts, all sharply pointed, and the middle and lateral ones with their ends notched or cleft.

(B) Early Blue Violet; Palmated Violet (*Viola palmata*) has slightly smaller blue flowers with bearded side petals.

The basal leaves are very variable in shape, ranging from heart-shaped with rounded teeth and an unbroken edge to palmately cleft ones with five or seven rounded lobes. Both of these Violets are common in dry ground, the former in fields or the borders of swamps, and the latter usually in thin woodland, from Me. to Minn. and southward.

(A) Common Violet (*Viola cucullata*) is the commonest and best known of all the Violets. It grows in low land everywhere—in woods, meadows, marshes, or along roadsides. It is a very beautiful and variable species both as to size and color of blossoms and to shape of the leaves.

The flowers are sometimes a deep purple and again may be a light blue, or even nearly white. The two upper petals are usually darker near the throat; the three lower ones shade to white at the throat, the side ones being beautifully fringed or bearded. The leaves are usually heart-shaped, round-toothed, and concave or furled; they are on long stems from the base.

(B) Canada Violet (*Viola canadensis*) is the most common of the leafy-stemmed blue Violets. You will notice that the preceding species all had their leaves from the base, and the flowers nodding on slender scapes, while this one has leaves growing on the slender stem and flowers above them on peduncles, springing from the angles of the leaves. This species is quite common in woods throughout the United States.

(A) Sweet White Violet (*Viola blanda*) is the most fragrant of our wild Violets, regardless of color. It is a most charming plant, but very diminutive, in fact, it is probably the smallest of the entire family. Occasionally we may find them in some exceptionally favorable locality growing to a height of perhaps 6 inches, but the usual height will barely exceed 2 inches. The plant is stemless, that is, the leaf stems and flower stalks all spring directly from the root.

The leaves of the common White Violet are rounded heart-shaped with slightly scalloped or round-toothed edges. It is very common in swamps and moist woods throughout the United States and southern Canada.

(B) Lance-leaved Violet (*Viola lanceolata*) is a taller, more slender species growing from 3 to 8 inches high. Its leaves are lance-shaped, scallop-edged, and on long stems from the root. The white flowers are only slightly fragrant; the three lower petals are strongly veined with purple and the two side ones are rarely bearded. It is commonly found in swamps and moist ground from N. S. to Minn. and southward, flowering from April to June.

Downy Yellow Violet (*Viola pubescens*) is a large, very handsome Violet that prefers, for its habitat, dry, hilly woods, often by the side of rushing brooks, but not usually where the soil is moist.

The Yellow Violet is one of the tallest members of the family, its stem ranging from 6 to 18 inches in length. Both the stems and the leaves are wooly-hairy. There are from two to four leaves growing from the stem near its summit; they are heart-shaped, pointed, and either toothed or scalloped. The flowers, rising on slender peduncles from the axils of the leaves, are rather large and bright yellow; the two lateral petals are heavily bearded and the lower one is handsomely veined with purple. These beards compel visiting insects to brush against the stigma and then against the anthers before reaching the nectar in the short spur.

Most of the Violets, during the summer, have apetalous or cleistogamous flowers on short peduncles from the root; these never open, but are fertilized in the bud. Common from N. S. to Manitoba and southward.

## LOOSESTRIFE FAMILY
## (*Lythraceæ*)

Purple or Spiked Loosestrife (*Lythrum Salicaria*) (European). Undoubtedly this species, which came to us from Europe, is the most beautiful of the genus.

The plant grows from 2 to 4 feet high and branches toward the top. The many purple flowers, making up the spike, each have six long petals and are trimorphous, that is, flowers on the same plant, have, relatively, three different lengths of stamens and pistils. Purple Loosestrife is found locally in swamps and on marshy borders of streams from Me. to Del. and westward.

Loosestrife (*Lythrum alatum*) is a tall, slender, native species growing 1 to 3 feet high, angular and branching. The deep-green, lance-shaped leaves are set oppositely on the lower stem and alternately on the upper branches. The flowers appear sparingly from the axils of the leaves near the ends of the branches. This species grows in moist ground from N. S. to Minn. and southward to the Gulf.

## MELASTOMA FAMILY
## (*Melastomaceæ*)

Meadow Beauty (*Rhexia virginica*) is a pretty little plant that always causes a thrill of admiration to pass through us as we come across it in grassy marshes where other flowers are usually few and far between.

It has numerous buds, two or three of which, only, open at a time, lasting but for a short space, the petals then falling off and the calyx and long stamens becoming withered and brownish; these detract greatly from an otherwise very beautiful plant.

Meadow Beauty or "Deer-grass" is a perennial, has a stout stem, quite branching and sharp-pointed, ovate, toothed, three-ribbed leaves, seated oppositely on the stem. The flowers grow on slender peduncles from the angles of the upper leaves; they have four large, rounded, magenta petals, each with a short, sharp point at the tip. The eight stamens are long and slightly unequal, the anthers being exceptionally large and bright golden-yellow. Meadow Beauty is found blooming during July and August in sandy marshes and shores from Me. to Fla. and in the states bordering the Mississippi.

## EVENING PRIMROSE FAMILY
## (*Onagraceæ*)

A family of herbs or shrubs with perfect, usually four-parted flowers, four petals, four sepals, four or eight stamens and a two- or four parted stigma.

(A) Great Willow Herb; Fireweed (*Epilobium angustifolium*) springs up in profusion and attains its greatest growth in clearings or recently burned land.

The tall, upright stem is usually simple, but occasionally slightly branched at the top. It attains heights of from 2 to 8 feet.

The flower spike is long; the flowers, blooming from the bottom upward, leave upright, long, slender pods.

The Great Willow Herb is abundant throughout our range in low ground, blooming during July and August.

(B) Hairy Willow Herb (*Epilobium hirsutum*) (European) has become naturalized and is fairly common in waste places and about old dwellings. It is branchy, hairy, has finely toothed, stemless leaves, and four-parted magenta flowers growing from the angles of the upper leaves.

(A) Evening Primrose (*Œnothera biennis*) is an exceedingly common biennial plant, of nocturnal habits, the flowers spreading wide open at dusk and partly or wholly closing the next morning.

The stem is soft-hairy, quite stout, and often very tall, ranging from 1 to 6 feet in height. Both the stem and the leaves are rather coarse in texture. The flowers are seated in the angles of the upper leaves. The four pale lemon-yellow petals are large and rounded, the flower spreading slightly less than two inches.

The lower buds open first, only a few at a time, so that usually we may find seed-pods seated among the leaves just below the flowers and undeveloped buds and leaves above. Primrose blooms in fields and roadsides, everywhere, from July to September.

(B) Sundrops (*Œnothera fruticosa*) a somewhat similar, diurnal species, with a branched stem, grows 1 to 3 feet high. The pale-yellow flowers measure from ½ to 1 inch across; they are in loose, terminal clusters or from the angles of the upper leaves. The leaves are linear-lanceolate, slightly toothed. Common from Me. to Minn. and southward.

## GINSENG FAMILY
## (*Araliaceæ*)

Wild Sarsaparilla (*Aralia nudicaulis*) has a single, large, compound leaf on a long stem from the creeping, fragrant, aromatic root. The flowers are gathered into three, rounded umbels at the top of a long stem that joins the leaf-stem near its base. Common in moist woodland from Newfoundland to Minn. and southward.

Ginseng (*Panax quinquefolium*) is well known as the plant that is collected and cultivated for its thick, fleshy, branching roots. The plant grows from 8 to 18 inches high. Three compound leaves, each consisting of five ovate-pointed, toothed, short-stemmed leaflets, radiate from near the top of the smooth stem. It is found in rich, cold woods from Quebec to Minn., southward.

Dwarf Ginseng (*Panax trifolium*) is a tiny species from 4 to 8 inches high. It has a spherical root, slender stem, three leaves compounded of three leaflets each, and numerous tiny white flowers in an umbel above them. Common in rich woods from N. S. to Minn. and southward.

## PARSLEY FAMILY
## (*Umbeliferæ*)

(A) Water Parsnip (*Sium circutæfolium*) is a stout, branching herb growing in shallow water. The rather weak stem is from 2 to 6 feet high. The alternating, compound leaves are very variable but usually of from seven to fifteen sharply toothed linear or lance-shaped leaflets. Flat, dome-shaped clusters, or umbels, of tiny white flowers terminate the upper branches. This species is very abundant throughout the country.

(B) Wild Carrot; Bird's Nest; Queen Anne's Lace (*Daucus Carota*) (European). While to flower lovers this may appear to be the most beautiful species of the family, it is the most heartily detested weed with which the farmer has to contend. It is very prolific, and each individual plant strikes its roots deep into the ground, as though determined to defy extermination. The fully-opened flower clusters have an exquisite, lace-like appearance, while those half-opened are hollowed suggestively like a bird's nest; in the centre of the cluster is a tiny purple floret, all the others being white.

## DOGWOOD FAMILY
## (*Cornaceæ*)

(A) Flowering Dogwood (*Cornus florida*) is a tall shrub or tree, ranging in height from 7 to 40 feet. The large, handsome flowers, 2 to 4 inches across, are in full bloom before or just as the leaves commence to appear.

The four large notched segments are not petals, but form the involucre and the real flowers are clustered at the centre; they have four tiny greenish-white petals and numerous little stamens. This Dogwood is common in dry woods from Me. to Minn. and south to the Gulf.

(B) Bunchberry; Dwarf Cornel (*Cornus canadensis*) is really a dwarf as compared to the preceding, for it grows only from 4 to 8 inches high.

The stem is leafless except at the top, at which point four to six leaves radiate.

What appears like a single large blossom seated almost within the whorl of leaves is in reality a cluster of tiny, green-petalled, four-parted flowers surrounded by four large greenish-white bracts. It ranges from Labrador to Alaska south to N. J., Ind., and Minn.

## HEATH FAMILY
## (*Ericaceæ*)

(A) Spotted Wintergreen (*Chimaphila maculata*) is a very handsome plant that we often come across in our rambles through rich woodland. The stalk, rising from 3 to 9 inches high, is of a ruddy color; the leaves are thick, smooth, irregularly toothed, lance-shaped, pointed, and with conspicuous whitish streaks following the veins. In July and August it bears one to five nodding flowers on long, erect peduncles above the topmost whorl of leaves. It ranges from Me., Ontario, and Minn. southward to Ga. and Miss.

(B) Pipsissewa; Prince's Pine (*Chimaphila umbellata*) grows in similar localities and is generally more common than the last. Its leaves are usually in two whorls about the brownish stem; they are bright shining green, toothed, unspotted, pointed, but broadened toward the end. The flowers are similar to the last and are in a loose 2- to 8-flowered umbel. The style is very short, with a five-parted gummy stigma. This species is found from N. S. to Ga. and westward.

(A) Shin-leaf (*Pyrola elliptica*) is the most common of the Pyrolas. The evergreen leaves are bright green, obscurely toothed, broadly elliptical and narrowing into long stems that clasp at the base. During May a long, smooth scape springs from the middle of the group of basal leaves to a height of 5 to 10 inches, bearing near its top a raceme of several flowers. It is common throughout the United States and southern Canada.

(B) One-flowered Pyrola (*Moneses uniflora*) externally closely resembles the preceding species. The flower scape is from 2 to 5 inches high, and at the summit bears, during June or July, a single nodding flower. It ranges from Labrador to Alaska and south to Pa. and Minn.

(C) Indian Pipe; Corpse Plant (*Monotropa uniflora*) is a very peculiar, ghostly appearing plant found commonly in dimly lighted rich woods. It has no green foliage, just white bract-like appendages on its upright, white, cold, clammy stem. A single white flower nods from the top. It is parasitic, drawing its nourishment from living roots or decaying vegetable matter. Common throughout our range.

Swamp Honeysuckle; White Azalea (*Rhododendron viscosum*) is a most beautiful swamp shrub with handsome, fragrant, white flowers. In low, wet swamps it is very common and blooms very profusely during June and July. The bush is from 3 to 8 feet in height and very branchy. The leaves are long-oval, broadest toward the blunt-pointed tip and narrowing to short stems.

The beautiful flowers are pure white, or rarely tinged with pink; the tube of the long corolla is covered with very sticky brownish hairs, and terminates in five, large-pointed spreading lobes. The stamens are very long, slender, and white, and tipped with yellow anthers. The five-pointed calyx is very small and inconspicuous.

During the early time of their bloom all the Azaleas bear hanging among the fragrant flowers, peculiar, juicy, pulpy growths that are edible, as any well-bred farmer's boy knows; he calls them May or Swamp Apples, but they are really modified buds and not fungous growths or caused by insects, as was formerly believed. These beautiful Azaleas are found from Me. to Ohio and southward.

(A) Pink Azalea; Pinxter Flower; Wild Honeysuckle (*Rhododendron nudiflorum*) is one of our most interesting wild shrubs, interesting because the flowers bloom before the leaves appear, or just as they commence to grow, and because of the very beautiful colors its pink flowers impart to our swamps during April and early May. The flowers are practically the same in form as the white varieties, except that the corolla tube is shorter.

Pink Azalea grows in open woods or swamps from Me. to Ill. and southward.

(B) Rhodora (*Rhododendron canadense*) is a beautiful member of this family, immortalized in verse by Emerson. It is a smaller shrub, growing from 1 to 3 feet high. The flowers usually appear before the pale-green, oblong leaves; the corolla is about one inch long, light magenta, and two-lipped. The upper lip is three-lobed and the lower is nearly divided into two distinct linear petals. They grow in thin clusters terminating the branches. Rhodora is found on damp hillsides and in swamps from Newfoundland to Quebec and south to N. J. and Pa., flowering during May and June.

American Rhododendron; Great Laurel (*Rhododendron maximum*) is a large, tall, and very ornamental shrub growing from 5 to 35 feet high. It is one of the most characteristic shrubs of the Alleghany Mountain region, where it grows in such profusion as to form almost impenetrable thickets. As it is a very hardy shrub and not injured by transplanting, it is very often used for decorative effects in parks and about private dwellings.

The oblong leaves are deep, glossy green, tough and leathery in texture, and have a smooth, slightly rolled-under edge. They droop in the winter season but are widespread in summer.

At the ends of the numerous branches, during June and July, are showy clusters of pink or white flowers. Each blossom spreads nearly two inches and is composed of five broad, blunt-ended petals of a pink-white color spotted with golden-orange. They have ten spreading stamens and a small pistil.

Rhododendron is found in rich, hilly, or mountainous woods, commonly from Pa. to Ga. but rarely northward to Ontario and Nova Scotia.

Mountain Laurel; Spoon-wood (*Kalmia latifolia*) is one of the most popular of our beautiful flowering shrubs. In the North it grows from 3 to 8 feet in height, but in the Southern States it often attains heights of 20 to 30 feet.

The leaves are dark, glossy green, pointed at each end and oblong in shape; they are arranged alternately along the branches and in dense terminal clusters. The flowers are very peculiar in their construction, the corolla being deep saucer- or bowl-shaped, with five short, broad lobes; on the outside, around the bottom edge of the "bowl," are ten small humps, that inside the corolla form little pockets to receive the anthers of the slender white stamens, curving from the centre of the blossom like the spokes of a wheel.

Both moths and bees visit these flowers in quest of the little supply of nectar that is secreted about the base of the greenish pistil. The flower stems are sticky so that only winged insects can get to the interior. Laurel is common from N. B. to Ont. and southward.

Sheep Laurel; Lambkill (*Kalmia angustifolia*) is a small, shrubby species, ranging from 8 to 36 inches high. Besides the common names given above, it is less often known as "Sheep Poison" and "Wicky," a rather sinister lot of names to be applied to a shrub with such handsome flowers.

All of the Laurels have dangerous properties, the juices of the leaves being very poisonous. It is also claimed that honey made by bees feeding on the nectar from Laurel blossoms is also poisonous. This species gets its many names, referring to its destructive effects on sheep, because it grows in abundance in pastures suitable only for the pasturage of sheep. The leaves of this small Laurel look tempting but are often very fatal to the animals eating them.

Their shapes, forms, and mechanisms are about like those of the Mountain Laurel, but the color is a beautiful, deep pink; little red anthers fit snugly in the ten little pockets formed for them in the surface of the corolla. Sheep Laurel is common from Lab. to Ont. and southward, blooming in June and July.

(A) Wintergreen; Checkerberry (*Gaultheria procumbens*). I doubt if there is a country boy or girl within the range of this plant, and it extends from Newfoundland to Manitoba and southward to the Gulf, who is not perfectly familiar with it.

The leaves are all clustered at the top of the ruddy stem that grows from 2 to 5 inches high; those of adult plants are deep, shining green, ovate-pointed, and very sparingly toothed. Usually two white tubular, 5-notched flowers hang on slender peduncles, just beneath the spreading leaves, during July and August.

(B) Trailing Arbutus; Mayflower (*Epigæa repens*). Arbutus is a creeping plant; the stems are tough, hairy, and branched; they spread out along the ground for 6 to 15 inches from the root. The evergreen, alternating leaves are tough, oval, slightly heart-shaped at the base, net-veined and toothless. The flowers are in terminal clusters, opening in April and May. They are five-parted, delicate pink, and have a fragrance similar to that of the Water Lily. Arbutus grows throughout the eastern half of our continent on shady, rocky hillsides.

(A) Labrador Tea (*Ledum grœnlandicum*) is an erect shrub growing from 1 to 3 feet high. It is not uncommon in suitable places in the eastern half of Canada, and is found rarely in mountains south to Conn., Pa., and Minn.; its habitat is in bogs or damp thickets. The narrowly oblong leaves are green above, have the edges rolled back, and are covered beneath with a rusty wool.

## DIAPENSIA FAMILY
## (*Diapensiaceæ*)

(B) Pyxie; Flowering Moss (*Pyxidanthera barbulata*). Pyxie is a very attractive moss-like shrub found commonly growing in the pine barrens of the Southeastern States, from N. J. to N. C. The branches are prostrate and creep along the ground for 6 to 10 inches from the roots.

Each branch is very thickly set with tiny, stiff, lance-shaped leaves less than half an inch in length. The numerous tiny white or pink flowers are seated on the stem, growing from the angles of the leaves; they have five broad-ended petals and attached between each of them is a curious, short, stout stamen.

## LEADWORT FAMILY
## (*Plumbaginaceæ*)

Seaside plants with perfect, regular flowers in one-sided racemes or spikes; five-parted and with plaited calyx.

Marsh Rosemary; Sea Lavender (*Limonium carolinianum*) is a very characteristic plant of the seashore; it is found very commonly in salt marshes along the Atlantic Coast from Labrador to Florida, and along the Gulf of Texas.

The plant has a thick, woody, very astringent root, from which grows a single naked stalk. This stem divides into numerous branches and branchlets, all destitute of leaves and spreading out so that the appearance of the whole plant is that of a very diminutive tree. The leaves all radiate from the root at the base of the flower stalk; they are spatulate-shaped, thick, almost smooth-edged, and are on long stems.

At the end of each branchlet is a slender one-sided raceme of tiny buds. From July until September these open out into tiny lavender flowers with five tiny petals, each coming from a five-toothed, ribbed calyx.

## PRIMROSE FAMILY
### *Primulaceæ*

(A) Yellow Loosestrife (*Lysimachia terrestris*). Yellow Loosestrife has a tall, slender, simple stem from 8 to 24 inches high. The leaves are pointed-lanceolate, stemless, and crowded along the stem, either oppositely or alternately. The flower spike is long and contains many buds on slender pedicels; they open from the bottom of the spike upward. Each flower has five-pointed golden-yellow petals, each with two small reddish-brown spots near the base; the stamens and pistil project in a cone-like cluster. This Loosestrife is abundant from Newfoundland to Hudson Bay and southward.

(B) Four-leaved Loosestrife (*Lysimachia quadrifolia*) is a very common species found in low land in about the same range. The flowers are very similar but each petal has a single large spot of reddish-brown at its base instead of a double one; the flowers appear from the axils of the upper leaves. The pointed, lanceolate leaves are whorled about the stem usually in groups of fours, occasionally more or less.

(A) Shooting Star; American Cowslip (*Dodecatheon Meadia*) is a western species that grows in open woodlands and on prairies from Pa. to Md. to Manitoba and southward through the Mississippi Valley.

The leaves are all in a tuft radiating from the base; they are oblong, bluntly pointed, and taper into trough-like stems. From the centre of this cluster of leaves rises a bare flower stalk, 8 to 20 inches tall, branching at the summit into several slender, curving peduncles, each supporting a single nodding flower.

The stamens project from the throat of the flower, the five golden anthers forming a conspicuous cone. Shooting Star blooms in April and May.

(B) Moneywort; Myrtle (*Lysimachia Nummularia*) (European) is a very dainty and beautiful trailing or creeping vine, often spreading over large surfaces of ground. It is a most beautiful plant for rockeries and does well in the house in hanging pots. The leaves, that grow oppositely all along the stem, are almost round; it is from their shape and the fact that they are about the size of the English twopence that they originally received the name of Moneywort.

Fringed Loosestrife (*Steironema ciliatum*). Fringed Loosestrife is a very branching herb not at all like the other varieties. The smooth stem rises to heights of from 12 to 24 inches. The species receives its specific name Fringed (*ciliatum*) because of the fine hairs on the upper side of the leaf stems, the rest of the plant being smooth.

The smooth light-green leaves are lance-shaped and pointed on short petioles or stems growing oppositely on the plant stem. The flowers grow on slender pedicles from the axils of the terminal leaves; the golden-yellow corolla is divided into five ovate lobes, each terminating in a sharp, twisted, or mucronate point; round the centre of the corolla is a reddish-brown ring, formed by the small spots at the bases of the five lobes. The pale-green pistil in the centre is surrounded by ten stamens, five being fertile and the other alternating ones being abortive.

Fringed Loosestrife is common in low ground and thickets from Newfoundland to British Columbia southward to the Gulf of Mexico.

(A) Star Flower (*Trientalis americana*) is a very dainty little plant often called the "Star Anemone."

The perennial rootstalk is long and horizontal and throws up a single, smooth, slender stalk from 3 to 9 inches high; at the top of this stalk is a whorl of from five to ten thin, smooth, veiny light-green leaves; they are lance-shaped and sharply pointed. During May and June a solitary blossom (rarely two) appears above the whorl of leaves on a very slender pedicel. The delicate white petals are sharply pointed and range from six to eight in number. The Star Flower is found in thin woodland from Labrador to Manitoba and south to Va., Ill., and Minn.

(B) Pimpernel; Poor Man's Weather-glass (*Anagallis arvensis*) (European) is a flower readily identified; in the first place there are very few red flowers to be found and no others with the shade of red of this one, a salmon or coppery-red. The square stem is smooth, slender, and rather weak, often lying prostrate on the ground. It is found in waste, sandy places especially near the coast.

## LOGANIA FAMILY
## (*Loganiaceæ*)

(A) Yellow False Jessamine (*Gelsemium sempervirens*). This beautiful vine is very common in the Southern States where it may be found climbing the trunks of trees, trailing over bushes, or even creeping over the ground. During March and April large, handsome yellow flowers appear in one-sided spikes. The blossoms are tubular-funnel form and range from 1 to 2 inches in length.

The ovate-pointed leaves, that grow oppositely on short petioles, are evergreen; the short flower spikes grow from their axils. The stem is smooth, woody, and twining. This species ranges from Va. to Fla. and Texas.

(B) Indian Pink; Pink-root (*Spigelia marilandica*) is an erect herb found in rich woods from Ohio and Ky. to Fla. and Texas. It is a perennial with a simple stem rising from 1 to 2 feet high. The flowers grow in a short, one-sided spike; the corolla is tubular-funnel form, five-lobed at the end, and about 1 or 2 inches in length.

## GENTIAN FAMILY
## (*Gentianaceæ*)

(A) Rose Pink (*Sabatia angularis*) is the most widely distributed of the Sabbatias. Whereas the rest of the tribe are confined in a range very close to the seacoast, this species is commonly found in rich ground in all the states from the Mississippi River to the Atlantic. Its period of bloom is during July and August.

The ovate-lanceolate leaves are stemless and seated oppositely on the stem. The branches usually divide near their ends, each division bearing a beautiful flower about an inch across. At the centre of the five-parted, pink corolla is a yellow-green star, a feature that is quite characteristic with members of this family.

(B) Sea Pink (*Sabatia stellaris*) is a beautiful, slender species common on salt marshes from Me. to Fla. The pink flowers grow singly at the ends of the slender branches. Like that of the last species, the centre is yellow-green but is often edged with a deep crimson which adds greatly to the attractiveness of the blossom.

Large Marsh Pink; Sabbatia (*Sabatia dodecandra*) is the largest-flowered and the most beautiful species of this genus; in fact, it is one of the most delicately beautiful of our wild flowers.

During July and August, along the Atlantic Coast, we sometimes find brackish ponds, the shores and muddy flats of which have a ruddy glow owing to the number of these large, attractive blossoms that appear. The stems are slender and wiry, and but little branched; they attain heights of 1 to 2 feet, each branch bearing usually but a single blossom.

The flowers measure from 2 to 2½ inches across; the nine to twelve petals are a delicate rose color and each has, at its base, a yellow-green spot margined by a three-pointed ochre or crimson border. The corolla has a regular, symmetrical wheel-like appearance, the petals making the spokes and the yellow centre forming the hub. The calyx is composed of linear sepals to the same number as the petals. The stamens are quite widely separated from the slender style, so that self-fertilization is hardly to be expected.

Fringed Gentian (*Gentiana crinita*), because of its exquisite beauty and comparative rarity, is one of the most highly prized of our wild flowers.

The stem is stout, stiff, and branching, each branch being erect and terminating in a bud. The yellow-green leaves are ovate-lanceolate, seated oppositely on the stem.

The calyx is angular, has four sharp points and is a bronze-green in color. During September and October we may find these blossoms fully expanded, delicate, vase-shaped creations with four spreading deeply fringed lobes bearing no resemblance in shape or form to any other American species. The color is a violet-blue, the color that is most attractive to bumblebees, and it is to these insects that the flower is indebted for the setting of its seed. The anthers mature before the stigma is developed so that self-fertilization is impossible. The flowers are wide open only during sunshine, furling in their peculiar twisted manner on cloudy days and at night. In moist woods from Me. to Minn. and southward.

(A) Downy Gentian (*Gentiana puberula*) is a handsome species springing from a perennial root, the simple, straight stem rising from 8 to 18 inches high; the stem is usually rough and slightly hairy. The light-green leaves are stiff and seated oppositely on the stem. The flowers are borne in terminal clusters or, sometimes, from the axils of the upper leaves; they are bell-shaped with five triangular, slightly spreading lobes. In color they are brilliant violet-blue.

Downy Gentian is common in dry fields and on prairies from Pa. to Ga. and west to Minn. and Mo., flowering during September and October.

(B) Solitary Gentian (*Gentiana Porphyrio*) is a pretty little species growing in moist places from southern N. J. to Fla. The simple, slender stem ranges in height from 6 to 15 inches and bears at the summit a solitary, erect, bell-shaped flower, of a light ultramarine blue color; the five spreading lobes are notched at their bases. The flower is very large compared to the stem and leaves of the plant it grows upon; the blossom measures from 2 to 2½ inches long, which is about the length of the linear leaves.

Closed Gentian; Bottle Gentian (*Gentiana Andrewsii*) is the most abundant of all Gentians. The flowers are as peculiar in their way as those of the Fringed are in theirs. It is remarkable because the five parts of the corolla never spread; the flower remains closed. The flowers are cross-fertilized by the common bumblebee. He knows there is a supply of nectar at the bottom of each blossom and he has the wits and the strength to get at it. Slowly but surely he is able to force the closed lobes apart until his body is half concealed in the "bottle", and he is able to reach the bottom. As he leaves the flower he is certain to scrape off quantities of pollen on his head and almost sure to leave some of it on the receptive stigma of the next flower visited.

The stem is smooth and simple; it grows from 1 to 2 feet high. The leaves are rather large ovate-pointed, and narrowed into very short, clasping stems. The flowers grow in terminal clusters, set in the axils of the last pairs of leaves. Closed Gentian grows in moist places, often along brooks, from Me. to Manitoba and southward.

## DOGBANE FAMILY
### (*Apocynaceæ*)

A small family composed chiefly of poisonous tropical plants usually with milky, acrid juices.

(A) Indian Hemp (*Apocynum cannabinum*) is a rather unattractive species with a smooth branching stem, rising from vertical roots to heights of 1 to 4 feet. The ovate-pointed leaves are closely crowded on the stalk oppositely to one another.

The small, five-parted, greenish-white flowers grow in terminal clusters. We find this species very abundant in dry fields and thickets throughout our range; it flowers from June to August.

(B) Spreading Dogbane (*Apocynum androsæmifolium*) is a much more attractive species than its relative just mentioned. It grows from 1 to 4 feet high, and has many long, spreading branches. The short-stemmed, broadly ovate-pointed, pale-green leaves grow oppositely, to the ends of the branches. The clusters of flowers terminating the branches are composed of pink, bell-shaped blossoms, having five pointed, recurved lobes.

## MILKWEED FAMILY
## (*Asclepiadaceæ*)

A family of stout-stemmed plants having milky juices and, usually, large opposite or whorled leaves. Each blossom has five tiny structures shaped like wishbones, with pollen masses on each end. They are so placed that the visiting bee or butterfly is pretty sure of getting one or more of its legs caught in the sharp angle at the apex and must, in order to get free, tear the tiny arrangement from its support. He then flies to the next plant with this dangling from his legs.

Butterfly-weed; Pleurisy-root; Orange Milk-weed (*Asclepias tuberosa*) is the most brilliantly colored species of the genus.

The stem of butterfly-weed is usually erect, from 1 to 3 feet high; it is rather rough and has but little of the milky juices so common to the other species. The leaves are pointed-oblong, very short-stemmed or seated oppositely. The beautiful orange flowers grow in flat-topped clusters or umbels at the summit of the plant. It is found from Mass. to Minn. and southward.

Common Milk-weed (*Asclepias syriaca*) is the most abundant and the best known of the Milk-weeds. It grows everywhere along roadsides, in fields and on the borders of woods. The rather stout stem rises from 2 to 5 feet high and has numerous, opposite, large, oblong, short-stemmed leaves of a yellow-green color. Both the leaves and the stem are finely hairy and both yield quantities of a thick, bitter, milky fluid if they are broken or bruised anywhere.

The flowers grow in rounded clusters, often in a pendent position, from the axils of the upper leaves. They are very fragrant and secrete an abundance of nectar.

In the fall the clusters of lilac-colored flowers have been replaced by large, rough-coated seed-pods that are completely filled with the silkiest of flossy substance attached to the numerous black seeds; finally the pod bursts and liberates the seeds, each floating away on the breeze, sometimes aviating for several miles before coming to earth.

(A) Poke Milk-weed (*Asclepias phytolaccoides*) is a tall species growing from 2 to 6 feet in height. The flowers composing its clusters are fewer in number than those of the common milk-weed but much larger and of a clear, ivory-white color. The flower stems are long and slender so that the entire cluster is in a nodding position, it being the only one of the genus in which all the flowers are pendent. Poke Milkweed is found, usually in dry situations, along the edges of woods or along roadsides, from Me. to Minn. and southward. It flowers from June until August.

(B) Whorled Milk-weed (*Asclepias verticillata*) is a very slender species, common in dry woods and on prairies in the South; found north to Mass. and Saskatchewan. The stem is slender, simple, and rises from 1 to 3 feet high. The narrow linear leaves have their margins rolled under; they grow in closely clustered whorls about the stem, usually quite erect. The numerous small, greenish-white flowers grow in a round cluster or umbel at the summit of the stem. It is a very dainty species, one not apt to be confused with any other member of the family.

## CONVOLVULUS FAMILY
## (*Convolvulaceæ*)

Hedge Bindweed; Wild Morning Glory (*Convolvulus sepium*) climbs gracefully over walls, through thickets, or twines its stem tightly about those of other plants or shrubs.

The large funnel-shaped blossoms grow singly on slender peduncles from the axils of the leaves. The flowers remain open only during sunshine and occasionally on bright, moonlight nights. It is very commonly found in moist ground along roadsides or the borders of woods or thickets, throughout our range and also in Europe.

Common Dodder (*Cuscuta Gronovii*) is a very common little parasitic plant found in moist, shady thickets or among the shrubs and plants bordering ponds or streams. It germinates its seeds in the ground, and the slender stem rises until it comes in contact with some living plant, when the root dies and the Dodder gets its nourishment from its host by means of numerous little suckers. It has no leaves; the stem is orange and the clusters of minute bell-shaped flowers are white.

## POLEMONIUM FAMILY
## (*Polemoniaceæ*)

(A) Downy Phlox (*Phlox pilosa*). Both the leaves and stem of this species are covered with fine, downy hairs; the sharply pointed calyx is also hairy and sticky. The plant stands from one to two feet in height. The narrow, lance-shaped leaves are rather closely alternated along the stem. The flowers are in a flat-topped cluster at the summit. The five lobes of the corolla are widespread and bluntly pointed.

The present species ranges from Conn. to Manitoba and southward, flowering during May and June in dry woods or on prairies.

(B) Moss Pink; Ground Pink (*Phlox subulata*) is a low, creeping species that spreads over sandy or rocky ground, forming compact masses resembling moss. The stem is very branchy but grows only a few inches high; the ends of the branches turn upward and terminate in clusters of flowers varying from crimson pink to white. Ground Pink is found from western N. E. to Mich. and southward.

## BORAGE FAMILY
## (Boraginaceæ)

(A) Forget-me-not (*Myosotis scorpioides*) (European). Forget-me-not is a common wild flower in Europe and Asia, and is in this country as an escape and fairly well established in Nova Scotia, New England, New York, and southward. The stem is rather stout but weak; it rises about a foot in height, and is smooth but the leaves are rough and hairy. The flowers are borne in one-sided curving terminal clusters. The five, broad, rounded petals are sky-blue with a yellow eye; the undeveloped buds are pink. There are several species of Myosotis, the present one having the largest and most beautiful flowers.

(B) Wild Comfrey (*Cynoglossum virginianum*) is a common, rough-stemmed perennial growing in deciduous woods from Me. to Mich. and southward. The tubular corolla is pale blue; it is set in a five-parted hairy calyx. The basal leaves are large and ovate; the stem ones clasp the flower stalk with somewhat heart-shaped bases.

(A) Blue-weed; Viper's Bugloss (*Echium vulgare*) (European). This peculiar plant is locally abundant in dry fields and waste places in the East. It is regarded as a pest and is a difficult one to get rid of.

The stem is light green spotted with purple; it grows erect from 1 to 3 feet high. The flowers grow on leafy spikes springing from the stem near the top. When the first flowers appear, in June, they are close to the stalk at the base of the rolled-up, leafy spike. As they continue to bloom the spike gradually straightens and the open flowers appear farther and farther from t he stem. The showy, tubular corolla is bright blue, and is exceeded in length by the long stamens and three-parted style; the buds are pink.

(B) Small Bugloss (*Lycopsis arvensis*) (European). This is a very rough, bristly stemmed species, also naturalized from Europe, and now found in waste places near dwellings, from Me. to Minn. and south to Va. The lanceolate leaves are seated on the stem; they diminish to the size of bracts and pass into the racemes of small, tubular, violet-blue flowers.

## VERVAIN FAMILY
## (*Verbenaceæ*)

Herbs with opposite leaves and perfect but usually irregular flowers, the tubular corollas spreading into two lips or four or five lobes.

Blue Vervain (*Verbena hastata*) is our most common example of the genus. It is a tall, slender, rank-growing plant reaching heights of 2 to 7 feet. The leaves are dark green short stemmed, lanceolate, sharply toothed, and grow oppositely on the stem.

At the top of the stem are numerous slender flower spikes, each branching from the stem and assuming a vertical position, in a regular order suggestive of candelabra. These slender spikes contain many buds, the lower of which open first. From July until the end of August we will find rings of purple flowers about the spikes, gradually drawing nearer the ends as the flowering season advances and leaving behind a long trail of purplish calyces. The tubular corolla has five spreading lobes, a slender pistil, and two pairs of stamens.

## MINT FAMILY
## (*Labiatæ*)

(A) Self-heal; Heal-all (*Prunella vulgaris*). Along roadsides, in fields, and on the borders of woods, everywhere throughout the country, we will find this familiar flower. The stem grows from 6 to 15 inches high and is topped with a cylindrical flower-head composed of many two-lipped, tubular, purple florets. But few of these bloom at a time, commencing at the bottom, and the flowering season extends from June to September. The leaves are sparingly toothed and seated oppositely on long stems. Usually several leaflets appear from their axils and sometimes smaller flower-heads from the axils of the upper ones.

(B) Skullcap (*Scutellaria integrifolia*) is one of the handsomest of the Skullcaps, the tubular, two-lipped flowers in the loose terminal spike each measuring about one inch in length. The downy stem rises from 6 to 24 inches high and is set oppositely with toothless, lance-shaped, round-ended leaves. It is found in dry ground from Mass. to Fla. and along the Gulf.

(A) Ground Ivy; Gill-over-the-ground (*Nepeta hederacea*). (European) is a beautiful little trailing mint that grows very profusely about country houses.

The leaves rise from the stem in pairs; they are round, with heart-shaped bases, the edge cut into rounded lobes, and their whole surface is downy and veiny. The pretty little purple flowers grow in small clusters from the axils of the leaves. The upper lip is erect and slightly notched; the lower one has three spreading lobes and is spotted with dark purple.

Ground Ivy is found in blossom from May to July throughout the eastern half of our country.

(B) Catnip (*Nepeta cataria*) (European) is a very common mint, introduced from Europe, the aromatic foliage of which has a very peculiar attraction for all members of the feline race. The plant has a stout, square hollow stem from 2 to 3 feet tall and is downy, as are the sage-green, toothed leaves. The lilac-white flowers are clustered on peduncles from the axils of the leaves. Catnip is common throughout our range.

(A) Motherwort (*Leonurus cardiaca*) (European) is a simple, erect-stemmed mint growing from 2 to 4 feet high. It has a very decorative effect, the leaves being large at the base of the stem and rapidly diminishing as they approach the top; the lower ones are quite long-stemmed and all are palmately slashed. The flowers grow in round clusters surrounding the stem at the axils of the leaves.

The numerous flowers composing these clusters have tiny, two-lipped, white, pink, or purple corollas and minute stamens. Both the stem and the leaves have a wooly texture and the former are strongly veined. Motherwort is commonly found about old country dwellings and along roadsides. We find it in bloom from June until August. It is a much more leafy species than most of the mints.

(B) Hedge Nettle; Wound-wort (*Stachys palustris*) is a tall mint (1 to 3 feet) with a downy-bristly stem and purple, tubular, two-lipped flowers in a terminal spike and from the axils of the upper leaves; lower lip streaked and spotted. Common in moist ground from N. S. to Manitoba and southward.

Oswego Tea; Bee Balm (*Monarda didyma*) is one of our most brilliantly colored wild flowers. It grows along the shady borders of woodland streams or pools where its vivid coloring is in strong contrast with the deep greens of the surrounding vegetation. The stem is hairy and rather rough; it attains heights of two feet or more. The short-stemmed, broad lance-shaped leaves are light green, sharply toothed, and rather thin.

The flowers grow in rounded terminal heads, composed of numerous long, tubular, scarlet florets. The upper lip is long, arched, pointed, and often notched at the tip; the lower lip is three-parted, the middle one being longer than the side ones.

Nectar, seated at the base of the long tube, can only be reached by long-tongued insects. Best adapted to it are bumblebees and certain of the butterflies. The Ruby-throated Hummingbird, too, attracted to this, his favorite color, often partakes of the sweets. From July until September this beautiful species blooms in suitable localities from Quebec to Manitoba and southward to Ga. and Mo.

## NIGHTSHADE FAMILY
## (*Solanaceæ*)

(A) Bittersweet; Nightshade (*Solanum Dulcamara*) (European), although an immigrant, is quite common in the eastern half of our country. It chooses for its habitat moist thickets or the edges of ponds where there are plenty of shrub to help support it, for this species has weak stems.

The dark-green leaves are variable in form; some are lobed, others have small lateral leaflets, and still others have another pair of still smaller leaflets on the leaf stem. The flowers hang in loose clusters on long peduncles from the axils of the leaves. This species blooms from June until September.

(B) Black Nightshade (*Solanum nigrum*) is a native species with a smooth, erect, branching stem 1 to 2 feet high. The long-stemmed ovate leaves have a wavy-lobed edge. The five-parted white flowers grow in few-flowered clusters from the leaf axils, the round berries are black when fully ripe, and are quite poisonous. This species is found throughout our range.

Purple Thorn Apple (*Datura Tatula*) is a large, ill-scented, rank-growing weed with a stout, smooth stem from 1 to 5 feet high. The long-stemmed leaves have very irregular, coarsely toothed outlines. The lavender-colored, trumpet-shaped flowers are about four inches long. The flaring corolla has five broad, sharply pointed lobes and is seated in a light-green, five-parted calyx about half its length. Usually the color of the corolla is more intense on the lobes and often shades to white toward the base of the tube. After flowering a large green fruit capsule about two inches long appears; it is ovoid in shape and armed with stout prickles. The entire plant has poisonous juices. It grows in waste ground, especially about barnyards, from Me. to Minn. and southward.

Thorn Apple; Jimson Weed (*Datura Stramonium*) also comes from across the water; it is very similar to the preceding, grows in the same places and in the same range. The flowers are white and the leaves are lighter green; the stem is also somewhat stouter.

## FIGWORT FAMILY
## (*Scrophulariaceæ*)

(A) Common or Great Mullein (*Verbascum Thapsus*) (European). This well-known plant is one of the most common sights along roadsides and in dry fields. Its long stalk rises from 2 to 7 feet above ground.

Mullein leaves are very soft, with fine white downy hairs; they have given to the plant a name very often applied—"Flannel Plant." The ones on the tall stalk are smaller and diminish in size to bracts as they reach the bottom of the long flower spike. From June until September these flowers open a few at a time and last but a day. The light-yellow corolla has five uneven, concaved lobes and five protruding stamens.

(B) Moth Mullein (*Verbascum Blattaria*) (European) has a tall, very slender stalk at the summit of which is a loose raceme. The flowers are large, have five petals, very prominent stamens, and orange anthers. The upper leaves are lance-shaped, the lower ones have the margins deeply cut, toothed, and notched. It is common from Me. to Ontario and southward.

(A) Blue Toadflax (*Linaria canadensis*). This is a very slender and dainty species, the stem attaining heights of 5 to 30 inches.

The little tubular flowers are violet-blue in color; the corolla is two-lipped, the upper one having two lobes and the lower one three; the latter is pouch-shaped and extends backward into a very slender spur. Blue Toadflax is commonly found in dry, sandy fields throughout the United States and southern Canada.

(B) Toadflax; Butter-and-eggs (*Linaria vulgaris*), although an immigrant, has extended its range from the Atlantic to the Pacific and southern Canada to the Gulf of Mexico. The stem is simple and from 6 to 30 inches high. The narrow alternating leaves are grayish-green, covered with a whitish bloom.

The tubular yellow flowers have two-lipped corollas, the upper ones of two lobes and the lower of three, the centre of one which extends into a large sac-like spur and has a protruding, pouting, orange palate that closes the throat of the blossom. This arrangement is designed for the bumblebee, whose weight on the lower lip opens the flower so he can get at the nectar, while it is tightly closed to pilfering ants.

(A) Turtle-head (*Chelone glabra*). This is a moisture-loving plant found in swamps.

The stem is stout, smooth, and erect, from 1 to 3 feet tall. The leaves are lance-shaped, stemmed, pointed, and toothed. The flowers are clustered in a short spike at the summit of the stem, the corolla is tubular, about an inch in length, and is white, tinged with pink. The upper lip is broad, arched, creased and notched in the middle; the lower lip is three-lobed and wooly-bearded in the throat. Turtle-head blooms from July until September and ranges from Newfoundland to Manitoba and southward.

(B) Pentstemon; Beard-tongue (*Pentstemon hirsutus*) has a straight, slender wooly stem that grows from 1 to 3 feet high. The leaves are light-green, lance-shaped, rough-edged, or minutely toothed, the upper ones seated oppositely on the stem and the lower ones with short petioles. The small magenta-white flowers are in panicled racemes. The trumpet-shaped corolla has two lobes to the upper lip and three on the lower, the throat nearly closed by a hairy palate on the lower lip. Me. to Wisconsin and southward.

Monkey Flower (*Mimulus ringens*) is a perennial with a smooth, square, hollow stem growing from 1 to 3 feet in height and branching considerably. The leaves, seated oppositely on the stem, are lance-shaped, pointed, and slightly toothed. The flowers are few in number and are on long, slender pedicels from the axils of the upper leaves. They open one or two at a time. The pale-purple flowers have two large lips, the upper divided into two lobes and the lower one into three, all broad and wavy. Four white stamens and a pistil nearly fill the throat, at the mouth of which are two bright orange-yellow spots.

A small store of nectar is secreted in the base of the flower tube. The double-yellow palate serves to close the entrance to the tube so that small useless insects may not be allowed to partake of the sweets within. When, however, the burly bumblebee alights upon the lower lip, his weight causes it to droop and allow easy access to its meagre supply of nectar.

Monkey Flower is found in wet places from N. B. to Manitoba and southward.

(A) American Brooklime (*Veronica americana*), the prettiest of the Speedwells or Veronicas, is a very frail plant.

The stem is stout, smooth, hollow, and quite weak; the lower part spreads over the ground and frequently takes root at the angles of the lower leaves. At intervals, branches rise to heights of 6 to 15 inches, bearing from the axils of the upper leaves small four-parted blue flowers in loose racemes. The light-blue petals have purple stripes and a white spot at the base.

Brooklime has a long season of bloom, being found in flower from May until September. It is common in moist ditches and along brooks or in swamps, from Newfoundland to Alaska and south to Va. and Mo.

(B) Common Speedwell (*Veronica officinalis*) is a popular little plant. The prostrate wooly stem is erect at the end and terminates in a raceme of pale-lavender, four-petalled flowers, the lower petal of which is conspicuously smaller than the other three, a common trait of this genus. Speedwell is quite common through the United States and southern Canada.

(A) Purple Gerardia (*Gerardia purpurea*) is a pretty little species that decorates low, moist, sandy fields and meadows with its beautiful purple-pink blossoms. The slender stem is quite branchy and averages about a foot in height, though it occasionally attains heights of 2 feet. From three to eight flowers, opening one at a time, grow along the ends of each branch. The corolla is broad and about 1 in. long, bright purplish pink, the mouth of the funnel spreading into five rounded lobes, spotted or downy within.

All the Gerardias and Foxgloves are quite parasitic, attaching their roots to those of other plants and getting part of their sustenance from them. This species is found chiefly along the coasts of the Atlantic, the Great Lakes, and the Gulf of Mexico. It blooms from August to October.

(B) Smooth False Foxglove (*Gerardia virginica*) has a smooth, branching stem from 2 to 6 feet high. The large, lemon-yellow flowers measure nearly 2 inches long by an inch broad. The plant grows from Me. to Minn. and southward and blooms during August and Sept.

Scarlet Painted-cup; Indian Paint Brush (*Castilleja coccinea*). This singular species is a parasite-that is, it fastens its roots upon those of other plants and takes their nourishment from them.

The slender, hollow, reddish, angular, and hairy stem grows from a tuft of smooth-edged, oblong leaves. The stem leaves are rather small and, the upper ones especially, have the ends three-lobed; those near and surrounding the flowers have their ends scarlet, as though they had been dipped in a pot of red paint. The flower's corolla is almost concealed in the two-lobed cylindrical calyx, the end of which is usually a brilliant scarlet. The corolla is irregular, greenish yellow, with a narrow upper lip and a three-lobed lower one. They have, set in the upper lip, four unequal stamens and a long pistil.

The Scarlet Painted-cup is found in low, sandy ground from Mass. to Manitoba and southward.

Wood Betony; Lousewort (*Pedicularis canadensis*) is a peculiar plant that we find in dry woods and thickets and often along roadsides.

The flowering stems are stout, hairy, and leafy; they rise to heights of 6 to 18 inches. The leaves are all fernlike in form; many of them rise on long, hairy stems from the roots and smaller ones alternate up the flower stalk. The flower spike is short and densely flowered and contains many small bract-like leaves among the tubular flowers. The corolla is composed of two lips, the upper one being arched and strongly curved or hooked at the tip. The upper lip varies from a yellowish green in freshly opened flowers to a dull reddish on the mature blossoms, this latter being the beefsteak color alluded to in one of its common names.

Wood Betony is found from Nova Scotia to Manitoba and southward. It is quite abundant throughout its range and its flowers may be found from early May into the latter part of July.

## BROOM-RAPE FAMILY
## (*Orobanchaceæ*)

(A) One-flowered Cancer-root; Broom-rape (*Orobanche uniflora*) is an attractive little parasite with a subterranean scaly stem, each branch sending up one to four very slender stalks from 3 to 6 inches high and bearing at the top a single blossom each.

Their color varies from a pale purple to a cream color and they average about three quarters of an inch in length. It is found in moist woods throughout the United States and southern Canada.

(B) Beech Drops; Cancer-root (*Epifagus virginiana*). This peculiar growth is found almost exclusively in beech woods.

The stem attains heights of 6 to 20 inches. At the ends of the branches are a number of curved, tubular flowers; these are stained a dull magenta.

Beech Drops attaches its roots to those of beech trees and gets all its sustenance from them. It blooms form August to Oct. and ranges from N. B. to Minn. and southward.

## BIGNONIA FAMILY
## (*Bignoniaceæ*)

Trumpet Creeper (*Tecoma radicans*) is an exceedingly beautiful woody vine having a southern disposition.

The stem grows from 20 to 40 feet long and is either prostrate or climbing. Sometimes it extends over the ground, climbing over the bushes that may be in its path, and again it may take an upward course and climb the trunks and branches of small trees. As it is a hardy plant it is often seen in cultivation and is used to decorate porches in the North.

The flowers are trumpet-shaped, red within and tawny or orange on the outside of the tube. They grow in terminal clusters of two to nine blossoms, each in a cup-shaped, two-parted calyx. The corolla is about 2½ inches long and flares into five rounded lobes. Four anther-bearing stamens and a pistil are in the upper part of the tube. The leaves grow oppositely on the stem and are each composed of 7 to 11 ovate, toothed leaflets. We find this vine from N. J. to Ia. and southward.

## PLANTAIN FAMILY
## (*Plantaginaceæ*)

(A) Common Plantain (*Plantago major*) is a very familiar weed about ill-kept dooryards. The leaves are large and spreading; broad-oblong, on long, troughed stems that radiate from the root.

The flower stalk rises to about the same height as the next, but the flower-head is very long. The tiny white flowers open in circles about this head, slowly making their way toward the top in their succession of bloom, which lasts from June until September.

(B) English Plantain (*Plantago lanceolata*). The leaves all radiate from the base; they are lanceolate, sharply pointed, and set on long, troughed stems.

The flower stem is stiff and smooth and attains heights of 6 to 18 inches. The head is short and studded with tiny, four-parted, dull-white flowers, with long, slender stamens. There are often perfect staminate and pistillate flowers on the same plant. It is now as abundant in all parts of our range as it is in its native European home.

## MADDER FAMILY
## (*Rublaceæ*)

(A) Bluets; Innocence (*Houstonia cærulea*). These are very dainty and beautiful little plants that decorate our fields profusely from April until July. The stems are very slender, about 3 to 6 inches tall, and have a few pairs of tiny leaves; larger leaves appear in tufts from the base. The perianth is slender and the lobes flare widely; the corolla is about one half inch in width, white, with the ends of the lobes pale blue or violet, and stained with yellow toward the centre of the flower.

(B) Partridgeberry (*Mitchella repens*) is a most beautiful little trailing vine with rounded, opposite, white-veined leaves along the creeping stem, that extends 6 to 12 inches from the root. Two beautiful little four-parted, bell-shaped flowers terminate each branch. They are downy white within, and pinkish and smooth on the outside. They have a fragrance similar to that of the Water Lily. A double red berry replaces the flowers in fall. It is common in woods throughout our range.

## HONEYSUCKLE FAMILY
## (*Caprifoliaceæ*)

(A) Bush Honeysuckle (*Lonicera canadensis*) is a bush or shrub with thin, straggling, brown branches, attaining heights of 2 to 4 feet. The leaves are thin, light green, somewhat heart-shaped and short stemmed. They grow oppositely on the branches and have small stipules between them. The flowers are borne in pairs from the axils of the terminal leaves. The Naples yellow tubes are about three fourths of an inch in length and have five lobes. This species is common from Quebec to Manitoba and south to Pa. and Mich.

(B) Twinflower (*Linnæa borealis americana*) is one of the most delicately beautiful of our wild flowers. The stem is slender, trailing, reddish brown, and from 6 to 24 inches long; at intervals very slender, leafy flower stalks rise, bearing at the end two pendulous, bell-shaped, white, fragrant blossoms; the corolla, which has five lobes, is crimson-pink within. The evergreen leaves are short stemmed, almost round, and scallop-toothed. Cool, mossy woods from Lab. to Minn.

Coral or Trumpet Honeysuckle (*Lonicera sempervirens*) is a very ornamental, climbing, woody vine growing from 8 to 15 feet in length. It trails over bushes or entwines its stem about the branches of trees. The lower leaves have short stems, are rounded-oval in shape, and opposite, as are those of all the members of this family. The leaves near the ends of the branches are united at their bases, clasping the stems and forming cup-shaped structures. The strikingly colored flowers grow in whorls on spikes terminating the branches. The tubular corollas are about two inches in length, bright red on the outside and yellow within; the opening of the corolla spreads but very little and is five-lobed. In the South the leaves of the Coral Honeysuckle are evergreen but in the North they are deciduous. In fall where each flower was located during the summer we find an orange-red berry. This species is distributed from Conn. and Neb. southward.

## BLUEBELL FAMILY
## (*Campanulaceæ*)

(A) Bellflower (*Campanula rapunculoides*) (European). This beautiful European species is a frequent escape from gardens and is quite firmly established in several localities in the Eastern States.

The simple stems are erect and quite tall, ranging from 1 to 3 feet high. The toothed, lance-shaped leaves alternate along the lower portion of the stem and the bell-shaped purplish flowers are in loose spikes on the terminal portions.

(B) Harebell; Bluebell (*Campanula rotundifolia*) is the "Blue Bells of Scotland" so familiar to us in song and verse. It is a very slender-stemmed species but very hardy, as attested by the altitudes at which it is found on mountains. The flowering stems are very slender and wiry, sparsely set with linear leaves; they usually branch near the summit, each division bearing a demure, drooping violet bell. It is found in bloom from June until September in rocky or sandy places in Canada and northern United States.

## LOBELIA FAMILY
## (*Lobeliaceæ*)

Cardinal Flower (*Lobelia cardinalis*). Although exceedingly bright colored, these flowers are rightly classed as among our most beautiful wild ones. As might be expected from their color, they are visited by and chiefly fertilized by the Ruby-throated Hummingbird.

The simple stem grows to heights of 2 to 4 feet, from perennial creeping rootstalks that often throw up new plants; the stalk is hollow and rather closely set with alternating, lanceshaped leaves, the lower ones stemmed and toothed, the upper ones clasping the stem and nearly smooth-edged. The showy flower-spike is loosely set with bright red flowers; the narrow, tubular corolla proceeds from a five-parted calyx, and ends in two lips, the upper having two erect, narrow lobes and the lower a broad three-cleft one, velvety scarlet; the five stamens are united in an erect tube. The Cardinal Flower is found in moist ground, especially along brooks, blooms in August and September and is found from N. S. to Minn. and southward.

(A) Spiked Lobelia (*Lobelia spicata*) is a small flowered species having a simple leafy stem from 1 to 4 feet in height. The leaves vary greatly in shape from lance-shaped to oblong, and decrease in size rapidly as they approach the flower spike. The small, pale blue-violet flowers are set in short smooth calyces. The upper lip of the corolla has two small lobes and the lower one is divided into three, larger, spreading ones. It is commonly found in dry, sandy soil from N. S. to Manitoba and southward.

(B) Indian Tobacco (*Lobelia inflata*) is the most common of the Lobelias; it is found growing everywhere in either sandy or moist soil, in woods or in fields. The alternating leaves are pointed-oval and sparingly wavy-toothed; the lower ones are quite large, while the upper ones are very small. The simple stem is stout and quite hairy; it grows from 1 to 2 feet in height. The little blue-violet flowers are barely one quarter inch long, each seated in a large, smooth inflated calyx.

The flower calyces enlarge after the corollas have withered away, and form round seed-pods.

## COMPOSITE FAMILY
## (*Compositæ*)

(A) Tall Blazing Star (*Liatris scariosa*) is a tall, handsome perennial that grows in dry situations and attains heights of 2 to 6 feet. A long spike containing numerous, quite large flower-heads adorns the top of the stem. These heads, which are about ¾ in. in diameter, have a very disheveled appearance for the magenta-purple rays emerge in all directions; they are contained in a large imbricated involucre. The leaves are stiff, lanceolate, and closely alternated along the stem. It is found from Me. to Mich. and southward.

(B) Ironweed (*Veronia noveboracensis*) is a tall (3 to 7 feet) and smooth-stemmed member of the Composite Family. The alternating leaves are lanceolate and finely toothed. The flower-heads are grouped in flat-topped clusters. The rays are slender and very numerous, giving the heads the appearance of little thistles. This species blooms in August and September, at which season it is one of the characteristic plants in moist ground near the seashore.

(A) Joe Pye Weed (*Eupatorium purpureum*) is a very familiar and pretty species. The simple, rather slender, stem is very tall, attaining heights of from 2 to 10 feet. The stem is usually stained purplish and is set at intervals with whorls of three to six, rough, coarsely toothed leaves; these latter have short stems, rather broad bases, and are sharp-pointed. The flowers grow in flat-topped terminal clusters. Each floret is of a rosy purple color that has projecting styles that give the flowers a very fuzzy appearance.

Joe Pye Weed is commonly found in moist places from Newfoundland to Minn. and southward, flowering during August and September.

(B) Thoroughwort; Boneset (*Eupatorium perfoliatum*) is a flowering herb, dearly beloved by the old-fashioned housewife and equally detested by the small boy. It was, and still is, one of the most commonly used home remedies. The stem is stout, hairy, and 1 to 5 feet tall. The opposite leaves are perfoliate, that is the ends are joined together. It is very common in swamps or thickets everywhere.

(A) Golden Aster (*Chrysopsis mariana*) is, as one would judge from its species name, very partial to the seacoast, where it may be found in profusion in dry sandy places and along roadsides.

The flowers grow in rather loose, flat-topped clusters, each head being on a rather long, slightly sticky peduncle. The tubular and ray florets proceed from a bell-shaped involucre composed of overlapping bracts. Its period of bloom is during August and September and it ranges from N. Y. and Pa. southward to Fla. and La.

(B) Curved-leaved Golden Aster (*Chrysopsis falcata*) is a very beautiful species with a silvery, wooly stem, 4 to 10 inches high, closely crowded with stiff, linear, downy, slightly recurved leaves. The golden-yellow flowers spread about an inch; the tubular florets in the centre have a brownish orange tinge but the numerous straps or ray-flowers are the brightest of orange-yellow. This species loves dry, sandy soil and is most abundant near the coast from Cape Cod to the pine barrens of New Jersey. It may be found in bloom from the latter part of July until September.

## GOLDEN-RODS
## Genus (*Solidago*)

The Genus Solidago is a very large one, comprising more than eighty species.

(A) Silver-rod; White Golden-rod (*Solidago bicolor*) bears the distinction of being the only one of our very numerous Golden-rods that does not have golden flowers. Those of this species are white or cream-colored. The stem is usually simple and attains heights of from 10 to 30 inches.

Silver-rod blooms during August and September on dry ground, frequently along roadsides or the edges of woods, from N. B. to Minn. and southward to the Gulf.

(B) Blue-stemmed Golden-rod (*Solidago cæsia*) blooms during September and October. The simple stem is closely set with lanceolate, toothed leaves, and from the axils of those on the upper half of the stem appear loose racemes of flowers. The heads are rather larger than those of most of the Golden-rods and have from three to five, comparatively long, golden rays surrounding the tubular florets.

(A) Early Golden-rod (*Solidago juncea*) is a very common species and one of the earliest to bloom, being found in flower from July until September. The flowers are in a large graceful cluster, composed of numerous racemes, at the summit of a tall, simple stem 2 to 4 feet high. The stem is smooth, angular, and usually a ruddy brown. The leaves are smooth, lance-shaped; the lower ones toothed, but the upper with nearly even edges. The golden-yellow flowers have eight to ten rays. Found in dry places from N. B. to Sask. and southward.

(B) Rough-stemmed Golden-rod (*Solidago rugosa*) is a very hairy species, rough to the touch. The stem attains heights of 1 to 7 feet. The flower racemes spread in a broad, pyramidal panicle. The leaves are feather-veined, quite hairy and coarsely toothed.

Canada Golden-rod (*Solidago canadensis*) is perhaps the most common and the handsomest of the genus. The flower cluster is very large and plume-like. The leaves are thin, narrowly lanceolate and finely toothed. The rather slender stem ascends to heights of 2 to 7 feet. The flower-heads are rather small.

Lance-leaved Golden-rod (*Solidago graminifolia*). This species differs greatly in appearance from the usual form of most of the Golden-rods.

The stem is simple, angular, and slightly rough; it ascends 2 to 5 feet and near the summit sends up many slender wiry, leafy branches supporting flat-topped flower clusters. The flowers are crowded closely together but are very small and rather dull-colored; they have 12 to 20 minute rays. The leaves are small and narrowly lanceolate; they have three to five ribs and are toothless but have a rough edge. It blooms from August until October, very commonly from N. S. to Sask. and southward to N. J. and Mo.

Showy Golden-rod (*Solidago speciosa*) is a large species, from 3 to 7 feet tall, with a stout simple stem and a magnificent, bright golden-yellow, plume-like head; the flowers are comparatively large and have usually five rays. Readily distinguished by its leaves, the lower ones rather large, contracting into a margined stem, gradually decreasing in size to small lance-shaped ones at the top of the stem.

## THE ASTERS
## Genus (*Aster*)

The members of this genus are exceedingly numerous in species. They are very variable; some have large flower-heads, others tiny ones; some are quite tall, others barely rise above the ground; some have few flowers on a plant while on others they are exceedingly numerous. Their colors are usually some shade of purple or white.

New England Aster (*Aster novæ-anglæ*) is one of the largest of the genus, its stem attaining heights of from 2 to 6 feet.

The stem is stout, branched, and rather rough. The leaves are soft, hairy, lance-shaped, and clasp the stem by a somewhat heart-shaped base. The flowers are in a broad corymb at the top of the stem. They are quite large, measuring about an inch across; the 30 to 40 narrow rays are of a purplish color, often quite bright.

This species is common from Me. to Minn. and southward, blooming from August to October, frequenting dry ground.

Smooth Aster (*Aster lævis*) is a handsome species having a smooth, stout stem, from 2 to 4 feet high. The flowers are in lovely terminal clusters, each blossom measuring about an inch across; they are usually light violet-blue in color, although color is very variable with all the so-called blue asters; each disk is surrounded by 15 to 30 rays. The leaves are nearly smooth-edged, lanceolate, clasping the stem with a distinct heart-shaped base. The Smooth Aster is abundant from Me. to Minn. and southward, growing in dry soil and blooming in September and October.

New York Aster (*Aster novi-belgii*) is one of the very commonest of the "blue asters." The stalk is slender, very branchy, and grows from 1 to 3 feet in height. The leaves are commonly narrowly lanceolate but are very variable; they slightly clasp the stem with their bases. The numerous flower-heads are a trifle more than an inch across, the yellowish centre being surrounded by 15 to 24 lilac or blue-violet rays. This species abounds from Newfoundland to Florida and perhaps west to the Miss. Valley. It blooms in September and October.

(A) Heart-leaved Aster (*Aster cordifolius*) is a common species, readily identified by the shape of its leaves that are, the lower ones especially, heart-shaped and on quite long, slender, ciliate petioles. The stalk is slender, branchy, and grows from 1 to 4 feet high. The flowers are numerous but comparatively small, about ⅝ inch across; they have brownish yellow centres and 10 to 20 lilac, or lighter colored rays. It is a very common species in thin woods and thickets, or along their edges. Found from N. B. to Minn. and southward, flowering in September and October.

(B) Panicled Aster (*Aster paniculatus*) is a very tall, branching, slender-stemmed species, commonly found in moist ground and on the borders of woods or copses. The smooth stalk attains heights of from 2 to 8 feet. At the ends of the branches are numerous flower-heads about the size of a nickel, loosely panicled. The leaves are long, lance-shaped, nearly smooth, obscurely, or not at all toothed, and dark green in color. This is one of the palest colored of the "blue asters," the flowers are very light violet and often white.

(A) Heath Aster (*Aster ericoides*) is a common white Aster from Me. to Minn. and southward. The plant grows 1 to 3 feet tall and has many branches, each having simple, many-flowered stems racemed along their outer ends. All the stems, even the flower peduncles, are set with tiny, heath-like, linear leaves the apparently different size between the flowers of this and the last species is because the scale is different.

(B) Many-flowered Aster (*Aster multiflorus*) has, as its name would lead one to think, very many flowers, but they are small, averaging less than ½ inch across. In fact, most of the white-flowered species do have smaller flowers than the blue ones, but what they lack in size they more than make up in numbers. The stem is slender but very branching, making a bush-like plant. Each branch is terminated by short, many-flowered racemes. The leaves are tiny, light green and linear, smooth-edged but rough to the touch, crowded along the branches to their tips. This is a common species from Mass. to Minn. and southward, growing in dry places everywhere and blooming from September to November.

(A) Small White Aster (*Aster vimineus*) is still another of the tiny, white-flowered Asters. It has a tall, branching stem from 2 to 5 feet high; the branches nearly all leave the main stalk in a horizontal position and the inflorescence is chiefly on one side of the flowering stems. It grows in moist soil from Me. to Minn. and southward.

(B) Starved Aster (*Aster lateriflorus*) is a much-branched, slightly hairy species, common in thickets and fields from N. S. to Ontario and southward. The leaves are lanceolate and taper to a point at each end. The ray florets are usually less in number than most of the other white species.

Daisy Fleabane (*Erigeron ramosus*) is a common aster-like species found blooming in fields from June until October. The stem is rough-hairy, and grows 1 or 2 feet tall. The small daisy-like flowers grow in a corymbed cluster at the top of the stem; they are about ½ inch across, have quite a broad disc of tubular, yellowish florets, and very numerous, narrow, ray florets; these rays range from 40 to 80 in number.

(A) *Aster umbellatus* is a common species of white Aster found growing in moist woodland or thickets. It has smooth, leafy, branching stems from 2 to 6 feet tall.

The numerous flower-heads are in compound flat-topped corymbs; the centre, or disc florets, are greenish yellow and are surrounded by a few white rays, usually less than a dozen. It is a common species throughout the northern parts of the United States.

(B) Sharp-leaved Wood Aster (*Aster acuminatus*) is a low-growing woodland Aster growing from 1 to 3 feet in height. The leaves are quite large, sharply pointed, sharply toothed, and short stemmed. A few alternate along the lower portions of the stem and a number are so close together as to appear whorled about the stem, just below the flowers. The flowers are few in number, on slender pedicels. They have few white rays and a rather brownish centre; the rays are long, narrow, often wavy and give the flower a spread of from 1 to 1½ inches. It is quite a common species in cool rich woods from Labrador to Ontario and south to Pa. It blooms during August and September.

(A) Pearly Everlasting (*Anaphalis margaritacea*) is the largest-flowered and the prettiest of the everlastings.

The stems are simple, quite stout, white-wooly, leafy, and 6 to 30 inches in height. The leaves are long and narrow, have a smooth edge, are grayish green above and wooly below, and narrow into clasping bases; they are closely set around the stem from the base to the flower cluster.

The flowers are in flat-topped clusters; each head is composed of many pearly-white, dry, overlapping scales that surround brownish-yellow, tubular florets. Staminate and pistillate flowers grow on different plants. This Everlasting is very common on dry hillsides, in woods, or on recently cleared land. Its range extends from N. S. to Manitoba and southward to S. C. and Mo. It is in full bloom from July until September.

(B) Sweet Everlasting (*Gnaphalium polycephalum*) has a wooly stem and wavy, lanceolate, wooly leaves. The pearly flower-heads are oval in shape; they do not expand until after they have matured. It is common in pastures everywhere.

(B) Rosin-weed; Compass Plant (*Silphium laciniatum*) is a large, showy-flowered plant found on the western prairies. It has a stout, rough, bristly stem that attains heights of from 3 to 10 feet. The stem grows from a perennial root. The large leaves are pinnately divided, each division being linear and cut-lobed.

The flower-heads are very large, measuring from 2 to 4 inches across. They are sessile or exceedingly short stemmed, seated along the upper portion of the stout stem. They are disposed to present their edges north and south. Compass Plant is found on prairies from Mich. to North Dakota and southward; it blooms from July until September.

(A) Prairie Dock (*Silphium terbinthinaceum pinnatifidium*) is rather an attractive plant that also grows on prairies and the edges of copses. The smooth, slender stem ascends 3 to 10 feet high and bears a loose panicle of large, yellow-rayed flower-heads. The leaves mostly come from the root and lower part of the stem; they are slender-petioled and deeply pinnatifid. Found from O. to Minn. and southward.

Elecampane (*Inula Helenium*) (European) is a tall, stout, beautiful member of the composite family that comes to us from the Old World.

The stout, smooth, usually unbranched stalk grows from 2 to 6 feet in height and is leafy throughout. At the summit of the stem is a single (or sometimes two) large flower set on a peduncle from the angle of the upper leaf. A smaller, flat, bract-like leaf appears just below the flower involucre. The head measures 2 or 3 inches across and has a broad disc of tubular, yellow florets, these turning tan color as they age. The yellow rays are numerous, but very narrow, usually set at different angles and with some vacant places so that the flower has a rather disheveled appearance.

The upper leaves usually clasp the plant stem, while the lower ones are on petioles. They are broad, thick-textured, toothed and pointed; the large, whitish veins show very prominently; the upper surface of the leaf is rough, yellowish green, while the lower is lighter and wooly.

(A) Robin's Plantain (*Erigeron pulchellus*) is one of the earliest members of the Composite Family to bloom. In fact, it is often known as the "Blue Spring Daisy," a name which is very appropriate for it, much more so than the one it commonly bears. The very fuzzy, light green, juicy stalk attains heights of from 10 to 24 inches. Most of the leaves are in a dense rosette at the base of the stalk; they are spatulate in shape, indistinctly toothed and hairy throughout. From one to nine flowers, an inch, or slightly more, broad are grouped at the top of the stem. It is common everywhere, blooming in May and June.

(B) Purple Cone Flower (*Brauneria purpurea*) is a showy western species bearing a single, large flower-head with a conical centre of purple disc florets and surrounded by many large, notched, magenta rays. The stiff, hairy stem rises 2 to 3 feet high. The leaves, also stiff-hairy, alternate along it; the upper ones are toothless and seated on the stem, while the lower ones are sharply toothed; they are five-ribbed and deep green in color. Rich soil, N. Y. to Mich. and southward.

Black-eyed Susan; Yellow Daisy; Cone-flower (*Rudbeckia hirta*) is a beautiful, large-flowered, tough-stemmed species that is commonly found in dry fields and pastures throughout the East, although it is, by nativity, a western species.

The stem is hairy, rough, very tough, and grows from 1 to 3 feet in height. Single, large flowers are borne at the summit of each stem.

The involucre is composed of two rows of leaf-like bracts that spread as the flower opens, the outer ones extending almost as widely as the rays. The conical, dark purple centre is composed of long, tubular florets that ripen in successive circles about the cone, making a fringe of yellow pollen on its surface. The orange-yellow rays are neutral, with neither stamens nor pistils.

The leaves, scattered alternately along the stem, are stiff and hairy. The upper ones are lanceolate and seated on the stem, the lower ones are broader toward the tip, rather spatulate shaped.

Tall Cone-flower (*Rudbeckia laciniata*) is a tall, lanky member of this genus, with an entirely different temperament from that of the Black-eyed Susan. No hot, sandy, or dusty fields for this, but the cool depths of moist thickets. As usual with vegetation in moist, rich soil, its growth is luxuriant. The smooth, branching stem ascends to heights of 3 to 10 feet and is leafy throughout. Ordinarily, the plant does not grow more than 5 feet in height; those that exceed this height might be termed giants of the species. The lower leaves are very large, are on long petioles and are cleft into five or seven divisions; the lower and middle stem leaves are usually three-parted while the upper ones, or at least the ones nearest the flowers, are small and elliptical.

Several large flower-heads terminate the branches; they measure from 2 to 4 inches across. The central disc is, at first, hemispherical and green but finally becomes elongated and brownish. The rays number six to twelve and are bright yellow in color. This species blooms from July until September and is found from Me. to Manitoba and southward.

Ten-petalled Sunflower (*Helianthus decapetalous*). This is a slender-stemmed, graceful, showy-flowered Sunflower, common in damp woods and on the borders of thickets, from Me., Quebec, and Minn. southward.

The branching stem grows from 2 to 5 feet tall; it is slightly hairy-rough on the upper portions but smooth below. The leaves are thin, rather rough-broad lance-shaped, short-stemmed and grow oppositely on the stem; they are all sharply saw-toothed. The showy flowers, growing on slender peduncles from the ends of the branches, are 2 to 3 inches across. Though often with ten rays, they just as frequently have any number from 8 to 15.

Common Sunflower (*Helianthus annuus*) is the common garden Sunflower that often has such enormous heads. The normal, wild plant is common from Minn. to Texas and westward. The flower-heads range from 3 to 6 inches in diameter; it is only the cultivated variety, produced from this, that has the mammoth heads we often see.

In its wild state the plant grows from 3 to 6 feet tall. Their period of bloom is from July until September.

Jerusalem Artichoke (*Helianthus tuberosus*) is, like the Common Sunflower, a valuable species, and one that is often cultivated because of its edible roots, these being tender and of good flavor; they are eaten raw or cooked. Their value as articles of food was first discovered by Indians and by them imparted to our early colonists. The name Jerusalem, in connection with this plant, is a corruption from the name applied to the species by Italians (Girasole Articocco), meaning sunflower artichoke.

It is a handsome plant, the stout, leafy, hairy stalk growing from 3 to 12 feet tall and being topped with several large showy flowers. The large, three-veined leaves are hairy and have toothed margins. They are chiefly set oppositely on the stem, although some of the upper ones may alternate. The several flower-heads are large, measuring up to 3 inches across. The central florets are greenish yellow and are surrounded by from 12 to 24 lone, golden-yellow rays.

This species is often also known as the Canada Potato and the Earth Apple. Its range extends from southern Canada southward nearly to the Gulf.

(A) Beggar-ticks; Stick-tight (*Bidens frondosa*) is a plant familiar, to their sorrow, to all who roam the woods and fields during fall. Who has not had the pleasant task of sitting down and, one by one, removing the little two-hooked, black seeds that hang so closely to clothing.

Beggar-ticks, in appearance, is an uninteresting weed common everywhere in moist ground or along roadsides. The stem is very branching and is from 1 to 8 feet tall. The leaves are compounded of three to five sharply toothed, lance-shaped leaflets. The flower-heads are composed of tubular brownish-yellow florets, sometimes with no surrounding rays and again with a few tiny ones.

(B) Larger Bur-marigold; Brook Sunflower (*Bidens lævis*) is a very attractive species while it is in flower, but later, after the little seeds have formed, it has the same disagreeable traits common to all the members of the genus. The flowers of this species are 1 to 2 in. across, having 8 or 10 large, yellow, neutral rays surrounding the dull-colored disc florets. The stem is slender and branching, the leaves lance-shaped and toothed. Common in swamps and along brooks.

(A) Common White Daisy; Ox-eye Daisy (*Chrysanthemum leucanthemum*) (European) is a naturalized, floral citizen. It is so common and has become so widespread that it is even better known than most of our common native flowers.

This Daisy needs no description. We have two very similar kinds differing in the shape of the leaves, one being more pinnatifid than the other. The one shown on the opposite page is the most common, a variety of *leucanthemum* called *pinnatifidum*. The other variety has the ends of the leaves rounded and finely toothed but not cut or slashed.

(B) Feverfew (*Chrysanthemum parthenium*) (European) is found in some places in the East as an escape from gardens. The stem grows from 1 to 2 feet tall and is quite branching. The flowers are grouped in clusters; they are much smaller than those of the last species and have a comparatively broader disc of yellow florets. The leaves are broad, deeply pinnatifid, and each division further toothed or cut. It is locally naturalized from Mass. to N. J. and westward. It blooms from June until September, the same as does the last species.

(A) Yarrow; Milfoil (*Achillea millefolium*) is one of the most common of our wayside weeds.

The stem is stout, gray-green, usually simple, or forking near the top. The leaves, alternating along and clasping the stem, are soft and feathery—deeply and finely bipinnatifid.

The flowers grow in very compact, flat-topped clusters at the top of the stem. Each flower-head has a centre of short, tubular, yellowish florets that turn brown or grayish as they grow old; they are surrounded by from four to six round, white rays.

Yarrow is a very hardy plant; we may find it thriving beside roads where the dust has killed nearly ever other living thing. Its leaves have a strong, not unpleasant, aromatic odor.

(B) Mayweed; Chamomile (*Anthemis Cotula*) (European) is also a common weed found by the wayside in company with the last species. The stem is very branchy, 8 to 20 inches high. The leaves are very finely divided. The strong, unpleasant odor of the foliage will at once correct the impression that it may be a Daisy.

(A) Sneezeweed (*Helenium autumnale*) is a beautiful rather odd plant that brightens meadows and swamps during August and September. The stem is rather stout, smooth, and branching; it ascends from 2 to 6 feet. Alternating along the stem are numerous ovate, pointed, sharply toothed, bright green leaves.

It is the blossoms that attract our attention for, besides being very handsome, they are unusual in form. The hemispherical centre is composed of closely packed tubular florets and is surrounded by a number of broad, toothed, golden-yellow rays; the heads have an expanse of 1 to 2 inches. Both the tubular and the yellow pistillate rays are fertile.

(B) Tansy; Bitter Buttons (*Tanacetum vulgare*) (European) is abundant everywhere about houses and along roads, from the Atlantic to the Rocky Mountains.

The foliage is very bitter and is the foundation of many an old-fashioned remedy. The flowers grow in flat-topped clusters and are composed of round discs, or "buttons," of tubular florets only. It is a species not to be mistaken; it has an appearance, an odor, and a taste of its own. It blooms from July until September.

(A) Golden Ragwort (*Senecio aureus*). In late March and early April, Ragwort shows simply a tuft of stemmed, heart-shaped leaves, resembling those of Violets. A little later a stem ascends from the perennial root; a slender, tough, angular, twisting stem that finally reaches heights of 1 to 3 feet. During May and June they carry at their summit a loose cluster of bright, orange-yellow flowers. Each flower is composed of but 8 to 12 narrow, orange-yellow rays, surrounding a central cluster of tubular florets of brownish orange.

The stem leaves are chiefly sessile, deeply cut or pinnatifid. Ragwort grows most abundantly and most luxuriantly in swamps or moist ground, but is also found in dry places or stony pastures. Its range extends from Newfoundland to Wisconsin and southward.

(B) Arnica (*Arnica mollis*). The slightly hairy stem grows from 1 to 2 feet tall. The basal leaves are long petioled but the stem ones are sessile and opposite, shallow-toothed. At the summit are one to nine flower-heads on slender peduncles. About the central disc are 10 to 14 yellow rays, each with three notches in their ends. Canada and the mountains of northern U. S.

Burdock (*Arctium minus*) (European) is a very common plant on waste ground, along roadsides and the edges of woods. The plant is often 4 feet or more high. The lower leaves are very large, often more than a foot in length, heart-shaped, deep green and finely veined above, grayish beneath because of the fine wool that covers the under surfaces. The upper leaves are smaller, more ovate in form, and less densely wooly on the undersides. The flower-heads grow in clusters at the ends of the branches. The involucre is almost spherical, composed of numerous bracts, each terminating in a sharp, hooked point. Tubular florets, only, are seated within this involucre; they are purple and white in color, and secrete an abundance of nectar on which account they are frequented by honey bees.

The present species adopts the policy of the Beggar-ticks, but instead of single seeds, it attaches the whole bur-like head by means of its numerous little hooks. They cling tenaciously to everything they touch; doubtless most of my readers recall massing these burs together to make castles, funny men, animals, etc.

Canada Thistle (*Cirsium arvense*) (European) is a small-flowered, perennial species that has strayed across the ocean and become a pernicious weed.

The stem is rather slender, branching, and grows from 1 to 3 feet in height. It grows from aperennial, creeping rootstalk, that is, as farmers have discovered, very difficult to eradicate from the soil. It grows in extensive colonies and, unless strenuous efforts are made to destroy them, they very soon take possession of a field to the exclusion of almost everything else.

The leaves, that grow alternately and closely together on the stem, are long, lance-shaped, deeply cut into sharply prickled lobes. Numerous flower-heads, about one inch across, terminate the branches. When in full bloom, the florets vary in color from rose-purple to white; the involucre is almost globular and covered with over-lapping bracts, each with a tiny, sharp, out-turned point.

All the thistles yield an abundance of nectar and are frequented by bees and butterflies.

Bull Thistle (*Cirsium lanceolatum*) is the thistle that we most often see in fields and pastures. It is one of the largest of the genus, its heads often measuring 3 inches across. The stem is stout and simple, and grows from 1 to 3 feet high; it is hairy and angular in section and grows from a biennial rootstalk.

The flower-heads are very large, 2 to 3½ inches across and usually solitary, although frequently two heads grow on the same stalk. The leaves are lance-shaped, green, clasping, rather hairy, pinnatifid and armed with short, stout prickles. Just below the flowers are several small bract-like leaves, also armed with sharp prickles. All this armor tends to discourage pilfering insects from crawling up the stem; should they persist and reach the large involucre, which is also armed, they will find that, in addition, it is slightly sticky, and presents an impenetrable barrier to their upward progress. This species is common from Me. to Del. and Pa. It blooms from July until September.

(A) Star Thistle (*Centaurea nigra, var. radiata*). The slender stein branches slightly and rises to heights of 1 to 2 feet, each branch bearing a solitary flower-head at the end. The flower-head has a round involucre of tawny, or dark brown, dry bracts; the florets are all tubular and rose-purple.

This species, which is introduced from Europe, grows in waste places and along roadsides from N. S. to Ontario and south to N. J. and Pa. It may be found in bloom from July until September.

(B) Chicory; Succory (*Cichorium intybus*) (European) has become thoroughly naturalized and is common in the eastern half of the United States, especially so near the coast.

The stem is stiff, tough, and angular in cross-section; it attains heights of from 1 to 3 feet. The leaves are long-lanceolate, dark gray-green and coarsely toothed. The flowers are very beautiful—a violet-blue, approaching a pure blue in color. There are at least two ranks of strap-shaped rays, the inner ones much shorter, all toothed at the ends. Succory blooms in dry situations from July until October.

(A) Fall Dandelion (*Leontodon autumnalis*) (European) is a small dandelion, naturalized from Europe and common in the Eastern States during fall, or from the latter part of July. The leaves, tufted at the base of the flower scape, are long and narrow and have blunt teeth. The flower scape is long and slender and usually forks near the summit, bearing two or three flower-heads, rarely only one; the scape attains heights of 7 to 18 inches. The flower stalk is not hollow like that of the common dandelion, but is solid. It grows in fields and along roadsides and is quite common from Newfoundland to Mich. and south to Pa.

(B) Dwarf Dandelion; Cynthia (*Krigia virginica*) is a tiny little plant as compared to the common dandelion. The leaves are all basal on rather long petioles; they are coarsely and sharply, or laciniately, toothed. Numerous unbranching, slender flower scapes rise from these tufts of basal leaves, each bearing at the summit a little golden-rayed flower resembling a dandelion.

Cynthia is a very common native species and is found blooming from April until July in dry fields, open wilds, or sandy soil, from southern Canada to the Gulf.

(A) Common Dandelion (*Taraxacum officinale*) although an immigrant to our land, has extended its range from the Atlantic to the Pacific, and is as well, or better known, as any other wild flower that we have. As every one knows its green, jagged leaves form a staple article of food and can be purchased in markets in spring at so much per peck. This species, with its large, flat, rosette of leaves and bright sunny flowers needs no description; it is well shown on the accompanying plate. All parts contain a bitter milky juice that exudes freely whenever the plant is broken.

The dandelion blooms most abundantly during the spring months but may be also found during every other month, even in winter. The name dandelion, of course, refers to the jagged edge of the leaves.

(B) Red-seeded Dandelion (*Taraxacum erythrospermum*) is a smaller species, also European, with more deeply cut leaves (pinnatifid) and with reddish-brown seeds, whereas those of the preceding species are usually olive-green. Common in dry fields from Me. to Pa. and westward to the Mississippi.

(A) Sow Thistle (*Sonchus oleraceus*) (European). This species is not a real thistle at all and the name "Sow" is applied rather as a term of derision, signifying spurious or worthless.

The stem is stout, smooth, grooved, hollow, and succulent; it attains heights of 1 to 6 feet. The leaves are shaped more like those of the dandelion than a thistle, but are armed with soft spikes. The small, thistle-like flower-heads are light yellow; they grow in loose clusters, terminating the branches.

(B) Wild Lettuce (*Lactuca canadensis*) is one of the rankest-growing of our native plants. The milky-juiced, branching, smooth stem ranges in heights from 3 to 10 feet. The leaves are all very angular, cut, toothed, and gouged in all manner of forms. Those near the base of the stem are very large, often attaining lengths of more than a foot. They become smaller and less deeply lobed as they mount the stem, the upper, small ones being almost entire edged. The small, yellow-rayed flowers are numerous but uninteresting. At maturity they are succeeded by silky beards of down, proceeding from the deep vase-like involucres.

(A) Rattlesnake-weed (*Hieracium venosum*) is commonly found in dry sandy places and in open woods. It can readily be recognized by the tuft of spatulate leaves spreading from the root, each leaf having strong veinings of purple.

A solitary stem, branching slightly at the top, grows from the centre of the tuft of leaves. The flower-heads are composed of bright, golden-yellow rays seated in a rather deep involucre; they resemble little dandelions but the rays are fewer in number. Rattlesnake-weed blooms from June until September and ranges from Me. to Minn. and southward to Ga.

(B) Canada Hawkweed (*Hieracium canadense*) is a smooth, slender, leafy-stemmed species. Besides the large, spreading, basal ones, the stem is alternately set with stemless, lance-shaped, sharply toothed, light green leaves. The flowers are practically like those of the last species. In October, they are replaced by little brownish globes of down. Canada Hawkweed is common on the borders of woods from Newfoundland to British Columbia south to N. J., Mich., and Oregon. It blooms from July until September.